REED'S STEAM ENGINEERING KNOWLEDGE FOR ENGINEERS

THOMAS D MORTON

CEng, FIMarE

Extra First Class Engineers' Certificate

W0234882

REEDS

LONDON · OXFORD · NEW YORK · NEW DELHI · SYDNEY

To my wife and daughter for their patience and help
TDM
South Shields, 1979

REEDS
Bloomsbury Publishing Plc
50 Bedford Square, London, WC1B 3DP, UK
29 Earlsfort Terrace, Dublin 2, Ireland

BLOOMSBURY, REEDS, and the Reeds logo are trademarks of Bloomsbury
Publishing Plc

First published in Great Britain by Thomas Reed Publications 1969
Second edition 1974
Third edition 1979
Republished by Adlard Coles Nautical 2003

A catalogue record for this book is available from the British Library

Library of Congress Cataloguing-in-Publication data has been applied for.

ISBN: PB: 978-1-4729-6881-4

6 8 10 9 7

Printed and bound in Great Britain by CPI Group (UK) Ltd, Croydon CR0 4YY

MIX
Paper | Supporting
responsible forestry
FSC
www.fsc.org   FSC® C013604

To find out more about our authors and books visit www.bloomsbury.com
and sign up for our newsletters

PREFACE

The object of this book is to prepare students for the First and Second Class Certificates of Competency in the subject of Steam Engineering Knowledge.

The text is intended to cover the ground work required for both examinations. The syllabus and principles involved are virtually the same for both examinations but questions set in the First Class require a more detailed answer.

The book is not to be considered as a close detail reference work but rather as a specific examination guide, in particular *almost* all the sketches are intended as direct applications to the examination requirements. If further knowledge from an interest aspect is required the student is advised to consult a specialist text book, *e.g.,* turbines, boilers, control theory, *etc.,* as the range of modern marine practice has superseded the times whereby all the subject can be accurately presented in one volume.

The best method of study is to read carefully through each chapter, practising sketchwork, and when the principles have been mastered to attempt the few examples at the end of the chapter. Finally, the miscellaneous questions at the end of the book should be worked through. The best preparation for any examination is the work on examples, this is difficult in the subject of Engineering Knowledge as no model answer is available, nor indeed any one text book to cover all the possible questions. As a guide it is suggested that the student finds his information first and then attempts each question in the book in turn, basing his answer on a good descriptive sketch and writing occupying about $1\frac{1}{2}$ sides of foolscap in $\frac{1}{2}$ hour.

All the miscellaneous specimen questions given at the end of the book are from D.O.T. examinations together with the test examples at the end of chapter one.

ACKNOWLEDGEMENTS

The author wishes to extend his thanks to the following for permission generously given to use photographs, diagrams and information, etc.:

Foster Wheeler Boilers Ltd
Controller HMSO
Institute of Marine Engineers
Combustion Engineering Inc
Thompson Cochran Co Ltd
Babcock and Wilcox (Operations) Ltd
Negretti and Zambra Ltd
Stal-Laval
GEC Turbine Generators Ltd

CONTENTS

Page

CHAPTER 1—Boilers

Scotch Boiler, construction, defects and repairs, testing and blowing down. Water tube boilers, general construction, drum arrangement. Header boilers. Foster Wheeler boilers ESD I, II and III. Radiant, tangentially fired and double evaporation boilers. Superheat contol and desuperheating. Refractory. Steam to steam generation. Combustion equipment. Boiler mountings, blow down, feed water regulators, safety valves. Water tube boiler examination. Soot blowing. Water washing. Water treatment, distillation plant, demineralisation plant, deaeration plant, condensate line treatment. Conductivity. Salinometer.

Chapter 2—Turbine Theory

Nozzle theory, convergent and convergent-divergent. Velocity compounding. Pressure compounding. Stage efficiencies. Velocity diagrams for impulse and reaction turbines. Theoretical constructions of impulse and reaction blades. Reaction principle. Gas turbine. Free piston gas generator.

Proportional band. Proportional plus integral. Overshoot. Proportional plus integral plus derivative. Control actions. Pneumatic compound controller. Control practice. Controller types. Typical control loops, automatic soot-blowing, automatic combustion control, bridge control for turbine machinery, etc.

Chapter 6—Feed Systems and Auxiliaries

Open Feed Systems. Condenser. Theoretical considerations. Performance. Leak test. Tube failure. Loss of vacuum. Feed filters. Feed heaters. Direct contact. Surface. Closed Feed Systems. Advantages. Controller. Regenerative condenser. Air ejector. Turbo-feed pump. Hydraulic balance. De-aerator. Related Equipment. Evaporators. Single and multiple effect. Flash evaporation. Feed treatment.

Miscellaneous Specimen Questions

CHAPTER 1

BOILERS

Various types of boilers are in use in merchant vessels, Scotch, watertube, vertical, package and others. To cover each and every boiler in detail would require a book in itself, hence in this chapter we have restricted the coverage to encompass as many of the questions asked about boilers in the examination as possible.

SCOTCH BOILER

A number of Scotch boilers are still in use today as main and auxiliary units and few if any are being manufactured. Only a brief description will be given, more details can be obtained from the companion volume No. 8 of the series.

Construction

Three types of construction have been used (1) all riveted (2) riveted with welding (3) all welded. The majority of the more modern Scotch boilers are mainly the hybrid variety of riveted and welded construction, since the all welded type is more expensive to produce, this is due to the requirements for Class I fusion welded pressure .vessels.

Material:

Plain low carbon open hearth steel of good quality having an ultimate tensile strength between 430 MN/m^2 to 540 MN/m^2 is used. Steel produced by the 'Kaldo' or 'L.D.', or any other oxygen process would be acceptable. Most steel producing plants are phasing out their Bessemer converters and open hearth furnaces, but some open hearth furnaces have been converted to oxygen blast.

All flanged plating and stays, etc., which have had their ends upset must be heated to 600° C and then allowed to cool slowly in order to stress relieve.

Furnaces:

Corrugated for strength, this also gives increased heating surface area. Not to be thicker than 22 mm (excess material thickness would give poor heat transfer and could possibly result in the excess material being burnt off).

Most Scotch boiler furnaces are suspension bulb type corrugation, since, for a given working pressure and furnace diameter, their thickness would be less than for other corrugation types. This gives better heat transfer and efficiency.

Furnaces must be arranged and fitted so as to ensure that furnace renewal can be carried out with minimum possible inconvenience. With this object in mind riveted furnaces have their flange, which is connected to the combustion chamber tube plate, so designed that by suitable manipulation the flange can pass through the opening in the boiler front plating.

Combustion Chamber

These are made up mainly of flat thin plating and hence have to be given support by means of stays, girders and tubes.

Boiler tubes, in addition to carrying gases from combustion chamber to boiler uptake, support the boiler front tube plate and the combustion chamber front plate.

Stays and stay tubes give support to the boiler back and front plating between the combustion chambers in addition to giving support to the combustion chamber plating.

Combustion chamber girders which support the top of the combustion chamber may be built up or welded types.

Combustion chamber bottom plating requires no support by means of stays, etc., since it is curved and is hence the ideal shape for withstanding pressure (see Fig. 1.1).

Boiler end and wrapper plating

The boiler end plating, *i.e.* front and back, is supported by means of the combustion chamber stays and tubes. In addition, main stays about 67 mm diam are provided. These are in two groups of three between the furnaces and in two or three rows with about 400 mm

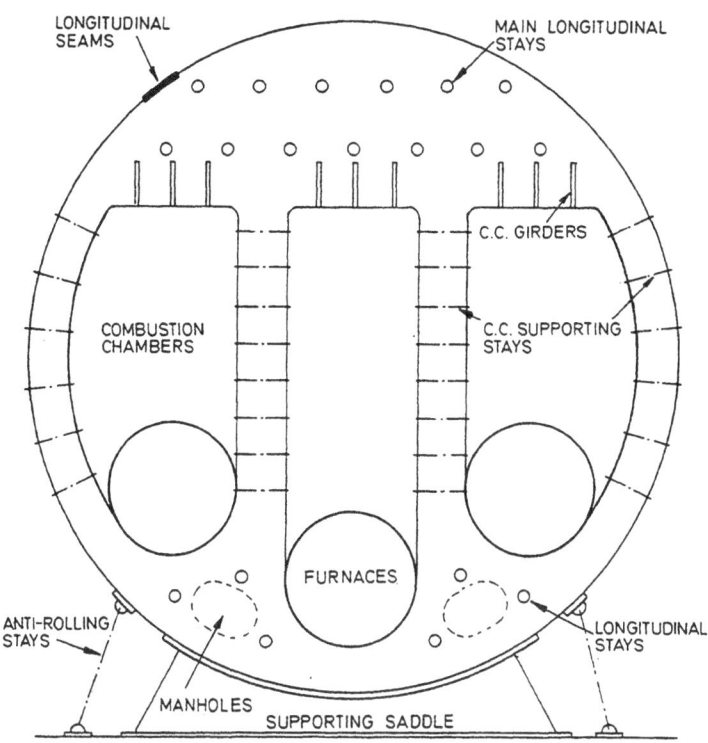

Fig. 1.1 SCOTCH BOILER

pitch at the top of the boiler.

The main wrapper plating is usually in two parts which are joined together by means of double strap butt joints of special design. Since the wrapper plating is circular it requires no stays for support against pressure.

Manhole openings are usually made in the bottom of the boiler back plate and in the top of the wrapper plating. These openings have to be compensated to restore the plate strength and the opening in the wrapper plating must have its minor axis parallel to the boiler axis, the reason for this is that the circumferential (or hoop) stress is twice as large as the longitudinal stress. [The student should revise the thin cylinder theory given in Part 'A' and prove $\sigma_H = PD/2t$ and $\sigma_L = PD/4t$]

Mountings

The disposition of the usual Scotch boiler mountings are as follows and points of importance relating to them are:

(a)   Water cock orifice for water gauge glass fitting is at least 127 mm above the top of the combustion chambers, and the top of the combustion chambers should be clearly marked on the outside of the boiler.

(b)   Internal feed pipe is closed at one end and perforated throughout its length in order to distribute the relatively cool feed water over a large space thus avoiding local overcooling of hot boiler plates and tubes.

(c)   Boiler blow down internal pipe goes to the bottom of the boiler in order that the boiler may be completely emptied. The blow down valve is either non return or a cock, usually the former to prevent back flow into the boiler of cold water if the boiler is blown down to the sea.

(d)   Scum cock is usually fitted in the steam space so that if the internal pipe becomes slightly or totally disconnected the boiler water level can not become lowered to a dangerous level inadvertently.

(e)   Main steam outlet is usually fitted with a dry pipe to reduce moisture carry over in the main steam supply (this is usual if the steam passes from the boiler to superheaters).

(f)   Two gauge glasses are fitted to determine the water level.

(g)   Whistle steam goes direct from the boiler (this also applies to a steam steering gear).

(h)   Connections are usually provided so that the boiler water can be circulated when raising steam.

When raising steam in a Scotch boiler care must be taken to ensure that the heating up process is uniform throughout the boiler otherwise straining of the boiler will take place and this could lead to tubes 'springing' and subsequent leakages, in addition cracks could develop in the plating if the stresses are high. It is customary to take as long as practicable to raise steam and this may be as long as 24 hours if no circulation means are provided other than natural convection currents.

(i)   Safety valves generally of the improved high lift type are provided to relieve excess boiler pressure, they are fitted with easing gear that usually comes down the side or front of the boiler to a convenient position. Drain lines but no cocks are provided and when the safety valves lift these drain lines should be checked to ensure they are clear, or damage to the valve chest may result.

Blowing Down and Opening Up a Scotch Boiler

If repairs or an examination of the boiler have to be carried out it will have to be emptied. It would always be better, if time is available, to allow the boiler to cool down in its own time after shut down, then pump the water out. In this way the relatively sudden shock cooling due to complete blow down would be avoided.

If the boiler has to be blown down to the sea, allow as much time as possible after shut down before commencing. The ships side blow down cock must be opened first then the blow down valve on the boiler can be *gradually* opened up. In this way the operator has some measure of control over the situation, if for example the external blow down pipe between boiler and ships side was in a corroded condition, then if the operator opened up the boiler blow down valve first, this could lead to rupturing of the blow down pipe and a possible accident resulting whilst he is engaged in opening up the ships side cock.

Our senses tell us when the blow down process is coming to a close, the noise level falls and the pressure will be observed to be low. Care must be taken to ensure that no cold sea water gets into the boiler, the boiler when empty of water would still contain steam which could condense and cause a vacuum condition, this in turn could assist the entry of cold sea water. To help prevent sea water entry, the boiler blow down is usually non-return (on some water tube boilers a double shut off is provided) but even with a non-return valve it is strongly advisable to start closing the boiler blow down valve when the pressure is low enough, and when it is down to the desired value, the valve must be closed down tightly and the ships side cock closed.

At this stage allow as much time as possible for the boiler to cool down and lose all its pressure, and when the pressure is atmospheric open up the air cock and gauge glass drains to *ensure* pressure inside boiler is atmospheric.

Either boiler door can be knocked in at this stage, top or bottom, but not both, provided sufficient care is taken. If it is the top door, secure a rope to the eyebolt normally provided and make the other end of the rope fast. Slacken back but *do not remove* the dog retaining nuts, take a relatively long plank of wood stand well back and knock the door down. The door is now open and the dogs can be completely removed, do not immediately open up the bottom door since if the boiler is hot this would lead to a current of relatively cool air passing through the boiler and subsequent thermal shock.

If it is the bottom door, slacken back on the dog retaining nuts by a very small amount, use a large plank of wood and break the door joint from a safe distance so that if there is any hot water remaining in the boiler no injury will occur to anyone. Again, do not immediately open up the top door of the boiler.

Hydraulic Test

When repairs have been carried out on a boiler it is customary to subject the boiler to a hydraulic test. Before testing, the boiler must be prepared. All equipment and foreign matter must be removed from the water space of the boiler and the repairs should be carefully examined.

Any welded repair should be struck repeatedly with a hammer to see if any faults develop, the sudden shock increases the stresses that may be in the weld and faults may then show up in the form of cracks.

The boiler safety valves have to be gagged and all boiler mountings, apart from the feed check valve and air cock, closed. The boiler can then be filled with clean water and purged of air. (Frequently the boiler is filled with water fed in through the top door by a canvas hose and when filled, the top door is fitted and boiler topped up through the feed line).

Using a hydraulic pump unit connected by a small bore pipe to the boiler direct or to the feed line, pressure can be gradually applied. The testing pressure is normally $1\frac{1}{2}$ times the working pressure, applied for at least 30 minutes.

With the boiler under pressure it can now be examined for leakages and faults. Weld repairs should again be given repeated blows with a hammer to see if they are sound.

WATER TUBE BOILERS

Water tube boilers have to a large extent superseded the Scotch boiler for the supply of steam to main and auxiliary machinery. Even donkey (*i.e.* auxiliary) boilers are frequently found to be water tube and certainly all modern turbine plants use them for main steam supply.

The advantages of water tube boilers are:

1. High efficiency (generally greater than 85%) hence reduced fuel consumption.
2. Flexibility of design—important space consideration.
3. Capable of high output (*i.e.* high evaporative rate).
4. High pressures and temperatures improve turbine plant efficiency.
5. Flexible in operation to meet fluctuating demands of the plant —superheat control rapidly responsive to changing demands.
6. Generally all surfaces are circular hence no supporting stays are required.
7. Steam can be raised rapidly from cold if the occasion demands (3 to 4 hours compared to 24 hours for a Scotch boiler) because of the positive circulation.
8. Considering a Scotch boiler and a water tube boiler with *similar evaporative rates* the water tube boiler would be compact and relatively light by comparison and its water content would be about 7 tonnes or less compared with the Scotch boiler's 30 tonnes.
9. With double casing radiation loss can be cut to 1% or less.

Various types of water tube boiler are now in use and envisaged, only some will be considered and this does not imply in any way that the boilers not considered are in any way inferior.

Construction
Materials
Drums. Good quality low carbon steel, the main constituents are 0.28% Carbon maximum, 0.5% Manganese approximately, 0.1% Silicon approximately, remainder mainly Ferrite. Ultimate tensile strength 430 to 490 MN/m^2 with about 20% elongation.

Steels with chrome, molybdenum, manganese and vanadium are increasingly being used. The increased strength and creep resistance enable less material to be used. Reduced weight, cost, machining and assembly time being advantages.

Superheater tubes. Plain low carbon (0.15% Carbon approximately) steel up to 400° C steam temperature. 0.5% Molybdenum low carbon steel up to 480° C steam temperature.

Austenitic stainless steel. 18% Nickel, 8% Chrome, stabilised against weld decay with niobium, for steam temperatures up to 590° C.

Weld decay: when stainless steel superheater tubes, in some earlier boilers, were welded to header stubs the microstructure of the metal adjacent to the weld changed. Corrosion protection by the chrome in the alloy steel was lost due to precipitation of the element as chromium carbide. A band of corrosion around the tube was named 'weld decay'.

Creep considerations predominate in the case of superheater tubes since they are subjected to the highest temperature (especially the last pass) and to boiler pressure. The actual metal temperature will depend upon (a) steam flow rate and temperature, (b) gas flow rate and temperature (c) tube thickness and material (d) condition of tube surfaces externally and internally. For normal conditions the temperature difference between inside and outside of the tube may be of the order, or less than, 38° C.

Creep tests are usually carried out over a period of 20 000 hours for superheater tubes in order to ascertain the creep rate and maximum strain. Creep rate would be approx. 10^{-6} m/mh and maximum strain 0.02.

For other boiler tubes, *i.e.* water tubes, the material used is generally plain low carbon steel since their operating temperature will be the saturation temperature corresponding to the boiler pressure.

Uncooled superheater element supports and baffles must have resistance to creep and corrosion. Alloys of nickel and chrome or steels containing high proportions of these elements are suitable.

Drum Construction

For steam and water drum, welding of preformed plating is the most usual method. Low pressure boilers have the steam drum plating uniform in thickness, with a single longitudinal welded seam. High pressure boilers may use two plates, tube plate and wrapper plate, with two longitudinal welded seams. The tube plate is thicker than the wrapper plate and it is machined to the thickness of the wrapper plate in the region of the weld (Fig. 1.2).

Test pieces made of the same material as the drum would be clamped to the drum and, using a machine welding process of the protected arc type, the weld metal would be continuously deposited on to drum and test pieces. When the longitudinal seam or seams are

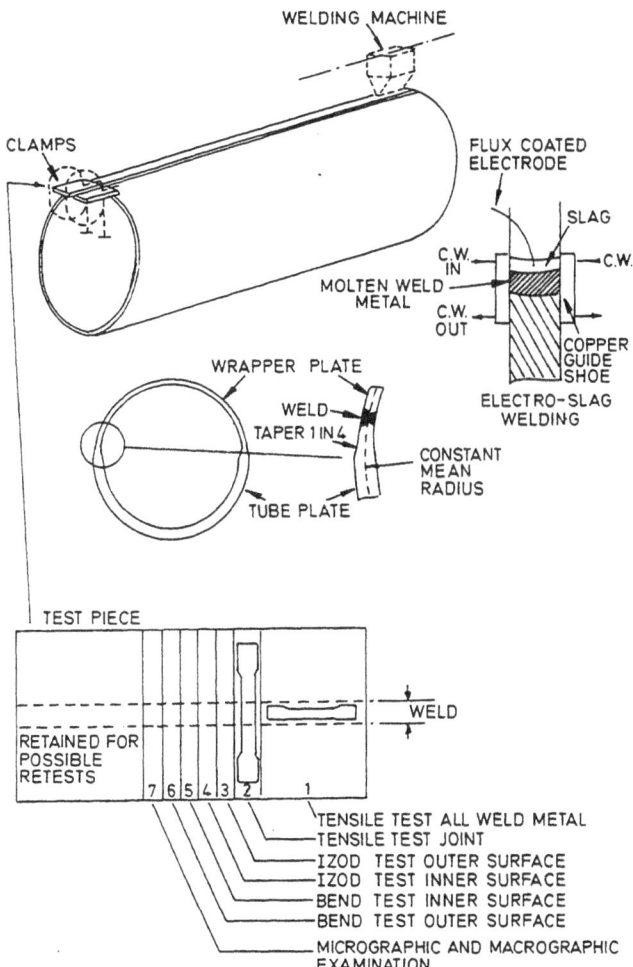

WELDING MACHINE

CLAMPS

FLUX COATED ELECTRODE

SLAG

C.W. IN
C.W.

MOLTEN WELD METAL
C.W. OUT

COPPER GUIDE SHOE

ELECTRO-SLAG WELDING

WRAPPER PLATE

WELD
TAPER 1 IN 4

CONSTANT MEAN RADIUS

TUBE PLATE

TEST PIECE

WELD

RETAINED FOR POSSIBLE RETESTS

7 6 5 4 3 2   1

TENSILE TEST ALL WELD METAL
TENSILE TEST JOINT
IZOD TEST OUTER SURFACE
IZOD TEST INNER SURFACE
BEND TEST INNER SURFACE
BEND TEST OUTER SURFACE
MICROGRAPHIC AND MACROGRAPHIC EXAMINATION

Fig. 1.2 LONGITUDINAL SEAM OF A WATER TUBE BOILER DRUM

completed, the test piece is then removed and the preformed drum ends would be welded into position. At this point, the welded seams, longitudinal and circumferential, are radiographed. The shadow pictures obtained will show up any defects such as porosity, slag inclusions and cracks, etc, these defects would then be made good by chiselling or grinding out and then welding.

Openings for boiler mountings, etc, would then be made and all the necessary fittings would be welded into position, *i.e.* branches, casing flanges, feet, etc. When all welding to the drum has been completed and radiographed the drum and test piece would then be annealed by heating slowly in a furnace up to about 600° C and then allowing it to cool down slowly.

The test piece is then cut up as shown in Fig. 1.2 and tested according to Class I welding regulations.

These regulations apply only to boilers whose working pressure is in excess of 4.44 bar and consist of:

(1)   Tensile test of the weld metal to check upon its strength and ductility.

(2)   Tensile test of parent and weld metal to check upon joint strength.

(3 and 4)   Izod tests to determine the materials notch brittleness and ability to withstand impact.

(5 and 6)   Bend tests to check ductility and soundness of material.

Tests 1 to 6 are well known and understood by engineers hence detailed descriptions are not warranted, however, macro and micro-examination require elucidation.

Electro slag welding of uniform thickness preformed plate to produce drums for high pressure boilers is being used. The drum is arranged vertically and the welding machine moves up a beam parallel to the seam. Fig. 1.2 shows simply a cross section through the seam and water cooled copper guide shoes. Main advantages of this welding process are (1) one weld run, this reduces possibility of inclusions, (2) up to 200 mm plate thickness can be welded in one run, (3) sound, reliable weld is produced.

Macro-examination

Preparation of test piece: this would consist of grinding and polishing until scratch free when viewed with the naked eye, then washing it in alcohol and then water ro remove grit and grease, etc. Next the test piece would be etched with an acid solution to remove the thin layer of amorphous (*i.e.* structureless) metal which will have been burnished over its surface.

Examination of the prepared test piece with the aid of a hand magnifying lens (x 10) may reveal cracks, porosity, weld structure and heating effects.

Micro-examination

Preparation of the test piece would be similar to that described above but the polishing process would be continued until the surface was scratch free when viewed under a microscope. After etching, the test piece would be examined under the microscope for defects. The pearlitic structure will be seen and so will any martensitic and troostitic structures, the latter two giving indication of hardening of the metal.

Different etching agents can be used, a typical one being NITAL which consists of 2 ml of Nitric acid and 98 ml of alcohol (methylated spirits).

Tube holes would now be drilled into the drums and the tubes fitted.

Tubes

These are arranged to form the furnace walls, etc, and to give positive circulation. Circulation is created by a force set up by the gravity head caused by differences in water density between tubes. This is effected by heat input, friction, head losses due to sudden contraction and enlargement and inertia loss.

Steam bubbles generated in tubes have lower density than the water, this gives natural circulation. However, difference in density between the steam and water decreases as pressure increases and there is no difference at critical pressure (220 bar). This causes problems for the boiler designer.

Water wall tubes frequently form the rear and side walls of the furnace and they may be fed with water from floor tubes which are supplied from the water drum, alternatively unheated large bore downcomer tubes external to the furnace may be used to supply the water wall tubes *via* their lower headers, no floor tubes being required. Often, water wall tubes have studs resistance welded to them in order to serve as retainers for plastic refractory. The advantages of water tubes are 1. Cooler furnace walls. 2. Reduced boiler size, since more heat is extracted per unit furnace volume. 3. Saving in refractory, initially, and because of the cooling effect, there will be less maintenance required.

Generating tubes are situated in the path of the furnace gases and are arranged to obtain as much of the radiant heat as possible in addition to baffling gas flow.

Return tubes, for water wall feeding and water drum feeding may be situated in the gas path in a lower temperature region or external to the furnace.

Superheater tubes may be situated in between generating and return tubes, in this way they create the necessary temperature difference to produce positive circulation in generating and return tubes, or they may be arranged in the gas uptakes after the generating section. The headers to which the superheater tubes are connected would be supply and return headers. Steam from the drum passes to the supply header, the steam then passes through the superheater tubes to the return header and thence to the main engine. Superheating methods are discussed on page 38.

Finally we have economiser tubes and air preheater tubes which are arranged in the flue gas uptake. In the case of the economiser tubes they may or may not be fitted with shrunk on cast iron gilled sleeves, the purpose of which, in addition to giving extra heating surface, is to protect the steel economiser tube from corrosion. The choice depends upon operating metal surface temperature.

Operating metal surface temperature depends upon, tube thickness, feed temperature, feed flow rate, gas flow rate and temperature. Generally if feed water inlet temperature to the economiser is 140° C or below, the sleeves will be fitted (Sulphuric acid vapour dew point is approximately 150° C or above, it depends upon various factors, the main one being fuel sulphur content). If the feed inlet temperature is above 140° C the economiser tubes may be of the extruded fin type.

Air preheater tubes made of plain low carbon steel generally have air passing through them and the gases around them, although the reverse arrangement has been used. These tubes are usually situated in the last gas path, hence they operate at the lowest temperature and are more likely to be attacked by corrosion products from the gases. Various methods have been used to reduce the effect of corrosion of these tubes, inserts have been provided in order to reduce heat transfer and corrosion over the first 0.3 m or so of the tubes, bypasses for the air used during manoeuvring to keep tube surface temperature high. Recently it has been suggested that glass tubes be used in place of the steel ones. Vitreous enamel has also been used as a coating for these tubes—unfortunately where it cracked off severe corrosion resulted.

Headers

For superheaters and water walls these are usually solid forged square or round section tubes with nozzles and ends welded on. Doors or plugs, opposite tube holes, may be provided to allow access for inspection, expanding or plugging of tubes.

The main methods used for securing tubes to headers are (1) expanding, (2) expanding followed by seal welding, (3) welding to nozzles or stubs.

Since welding on site can prove difficult (superheater elements may have to be renewed during the boiler's working life) it is usual to arrange for the number of site welds to be kept to a minimum with best possible access.

Fig. 1.3 shows various methods of securing superheater tubes:

(a)   Used if access to both sides of the weld is possible.

(b)   Welding from one side only.

(c)   This shows a five tube element welded to two 90° stubs on each header.

(d)   Tubes expanded into header. Access doors must be provided, this method is not used at temperatures above 470° C.

Babcock and Wilcox Header Boilers

This simple rugged boiler is not suitable for high pressure steam generation in large quantities which is the requirement of the modern turbine installation.

The advantages of the boiler are:

1.   Compact, space saving and robust.

2.   Easily and cheaply repaired.

3.   Easily cleaned.

4.   Reliable and simple to operate.

Construction

Fig. 1.4 shows simply the arrangement of a single pass header boiler suitable for auxiliary services. Single pass means that the gases pass straight up through the boiler from the furnace to the uptake, the older three pass boiler is fitted with baffles which tend to collect

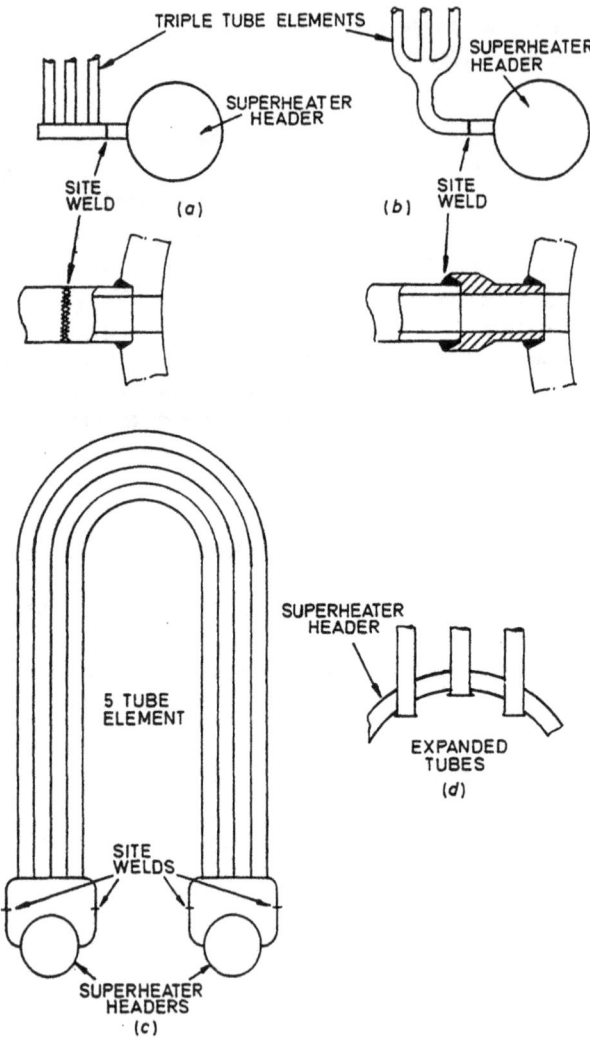

Fig. 1.3 SUPERHEATER TUBE SECURING METHODS

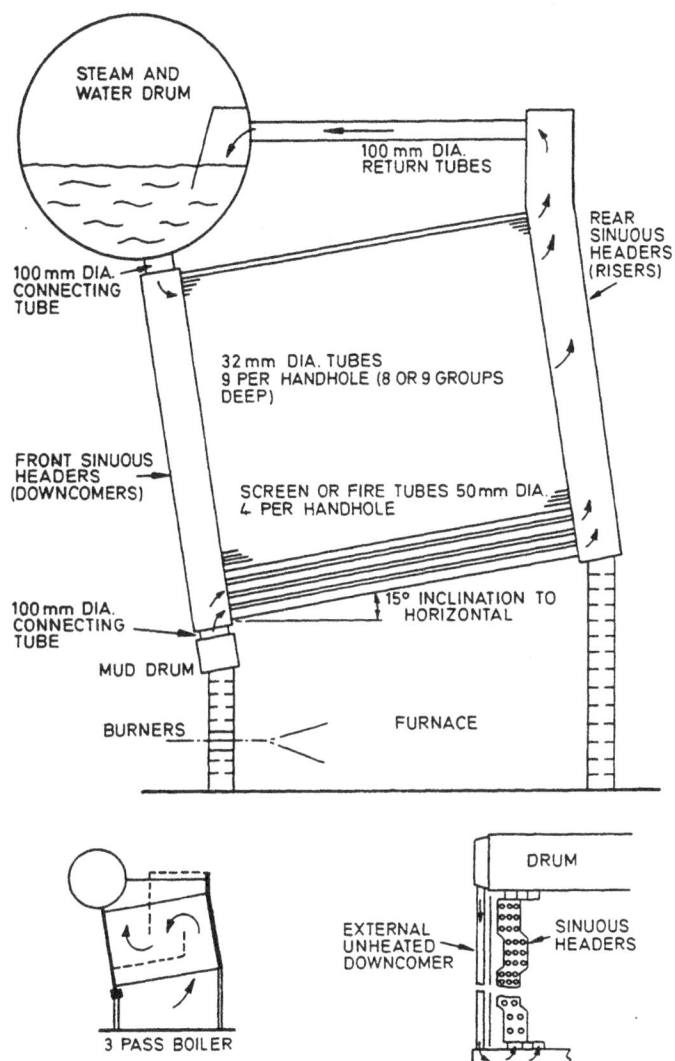

Fig. 1.4 HEADER BOILER

sooty deposits and can result in tube failure apart from affecting boiler efficiency. The straight generating tubes are expanded into sinuous headers. Headers are made sinuous to increase gas path, heat extracted from gases and reduce boiler height. Generating tubes are grouped into nine 32 mm diameter tubes per handhole and four 51 mm diameter tubes per handhole, eight or nine groups of 32 mm diameter tubes would be found in each header and two groups of 51 mm diameter tubes, the latter are termed screen or fire tubes since they are the tubes nearest to the furnace and therefore receive the most radiant heat from the flames (other tube sizes and groupings are possible).

Each front header is connected to the steam drum and sediment drum (mud drum or header) by means of short lengths of 100 mm diameter tube, the rear headers are connected to the steam drum by 100 mm diameter return tubes. Headers and tubes are inclined in order to improve circulation and steam generation rate.

In header boilers which are used for main propelling machinery the furnace walls will be filled with studded water tubes and super-heaters will be provided. The superheater U tubes are attached to two headers at the side of the boiler and the tubes occupy space in the generation bank that would be used by generation tubes in the saturated boiler, the front and rear headers are undrilled in way of the superheater.

Fig. 1.5 shows the superheater arrangement for a single pass boiler operating up to 48 bar, 480° C and 27 250 kg/h. Control of steam temperature is achieved by regulating the amount of steam passing through the submerged attemperator coil in the steam drum. Steam passes from the drum through the first superheater section to line A. If the regulating valve closes the by-pass line B, then all the steam passes through the attemperator coil to line C and then through the second superheater section. If the regulating valve opens up fully the by-pass line B, then steam passes from A to B and C by-passing the attemperator coil. In either case *all* the steam passes through the superheater sections thus ensuring no overheating due to steam starvation.

An alternative means of steam temperature control is to use gas dampers and a three pass system. With the arrangement, most of the gases can be made to either pass over the superheat section or by-pass the superheat section as the conditions demand.

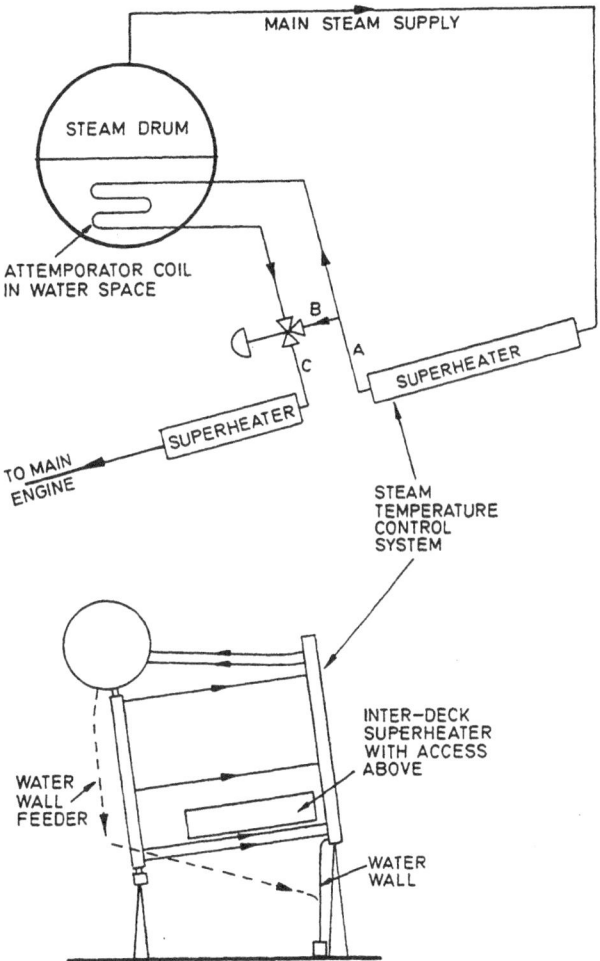

Fig. 1.5 SINGLE PASS HEADER BOILER WITH SUPERHEAT AND WATER COOLED FURNACE

Foster Wheeler D Type Boilers

D type boilers, the naming letter refers to the shape, *i.e.* like a letter D, are mainly in two groups. Those simply designated 'D type' are the older variety which operate at 32 bar steam temperature 400° C, up to 18 000 kg/h evaporative rate, and the E.S.D. type, *i.e.* external super-heater D type operating at 53 bar, steam temperature 515° C and up to 90 000 kg/h evaporative rate. Only the E.S.D. types will be considered in any detail.

Construction

Fig 1.6 shows the arrangement of the E.S.D. boiler, and for comparison purposes the older D type with internal superheater situated between the generating tubes.

The side, rear and roof of the furnace is basically made up of close pitched water wall tubes expanded into square section headers which are fitted with taper lip plugs for inspection and tube expanding, etc. Lower side and rear headers are fed from the water drum through the

Foster Wheeler E.S.D. 1 Marine Boiler

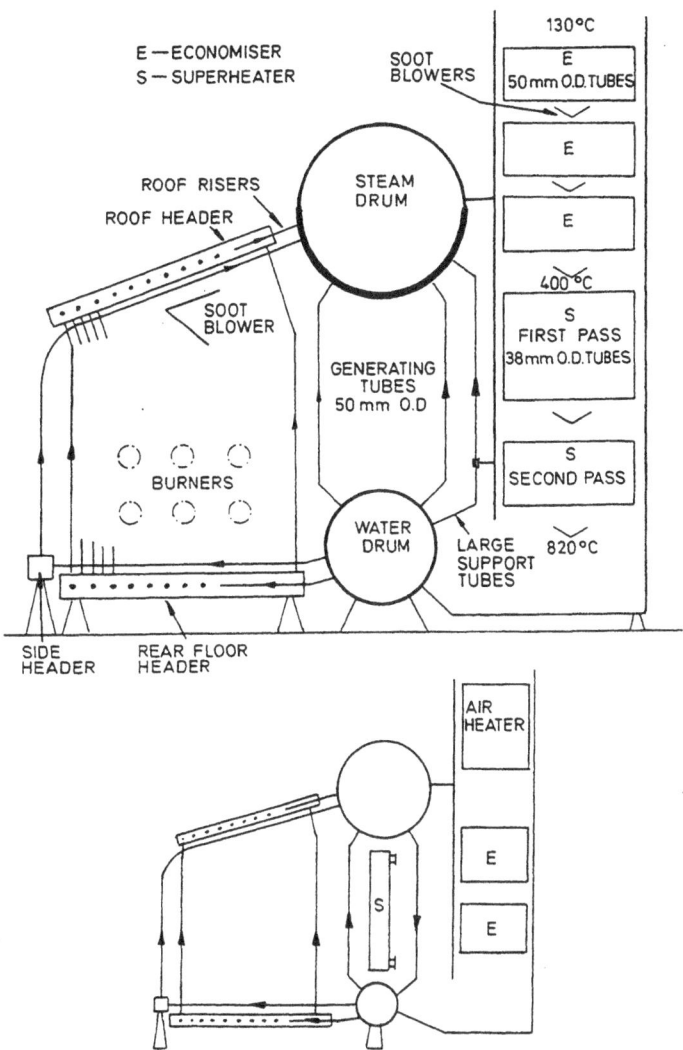

Fig. 1.6 ESD AND D TYPE BOILER

floor tubes, one floor tube supplies two water wall tubes. All generating tubes are risers and the water drum is fed by external large bore downcomers to ensure positive circulation (in the older D type about five of the thirteen rows of generating tubes after the superheater were risers the remainder downcomers under normal steaming conditions), about five downcomers would be found at front and back of the boiler, they are not shown in the figure. Large bore superheater support tubes, about six in number, are also to be found in the generating section, these are smaller in diameter than the downcomers.

Superheaters are situated after the generating section but before the economiser. They consist of multi loop elements generally welded to tube stubs which are welded to the circular headers (the headers are supplied with the tube stub welded in place and the element is welded to these during boiler erection) that are situated at the back of the boiler. Superheater element support beams cast from 18% Chrome, 55% Nickel heat resisting steel are fitted to the casing above the upper superheater section, and to large bore support tubes and the casing for the lower super-heater section.

Fig. 1.7 gives details of the welded type of superheater fitted in the lower portion of the boiler uptake, allowances have to be made for the expansion of the support beams in addition to the allowances indicated for the elements.

Economisers are built up with steel tubes fitted with shrunk on cast iron gills to give protection and extended surfaces.

Steam temperature control for the E.S.D. boiler is achieved by fitting an air attemperator between the first and second passes of the superheater. The attemperator is made of steel tubes with mild steel gills fitted to the tubes to give an extended surface, air flow across the attemperator is controlled by means of linked dampers. A bled steam air heater is fitted in the air ducting before the linked dampers to give additional heat to the air which is supplied from the forced draught fan to the burners, the complete arrangement of air heater, dampers, fan and attemperator is situated at the boiler front.

E.S.D. II Boiler

This boiler was first installed into vessels in 1961 and is an improved version of the E.S.D. type. It operates at 56 bar 490° C steam temperature.

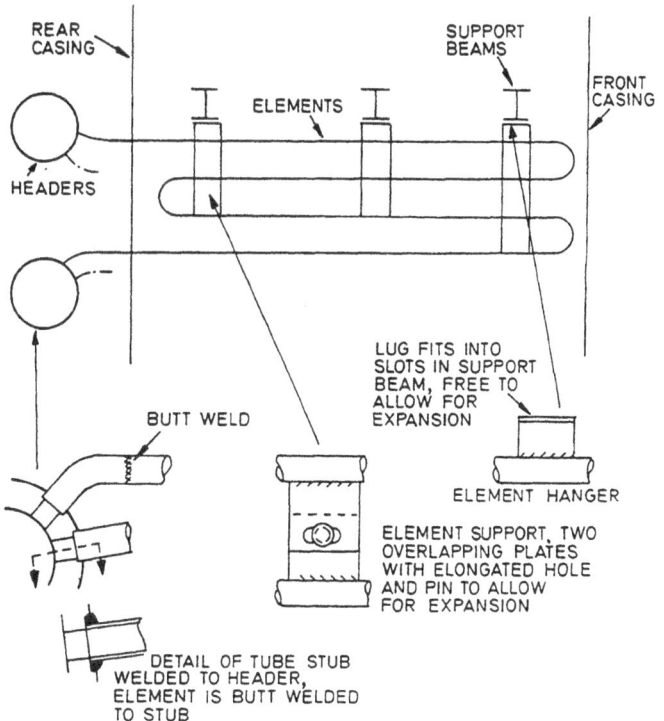

Fig. 1.7 WELDED SUPERHEATER

The main differences in construction and operation between the two boilers are:

1.   Floor tubes are usually slightly inclined to improve circulation.
2.   Increased space at boiler front as air attemperator and the bulky by-pass with linked dampers are dispensed with.
3.   Combustion air used for casing cooling.
4.   Wider scope for superheat control.
5.   Reduced steam pressure drop from drum to superheater outlet.
6.   Superheat temperature control is by gas dampers.

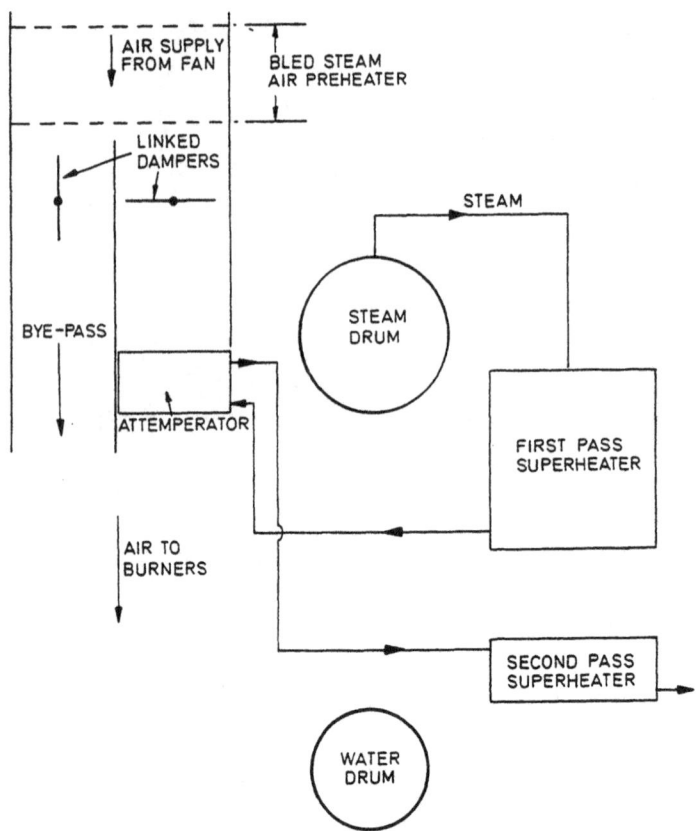

Fig. 1.8 STEAM TEMPERATURE CONTROL SYSTEM

Fig. 1.10 is the basic design of the E.S.D. II boiler shown in diagrammatic form for examination purposes.

Steam temperature control for the E.S.D. II boiler is by means of gas dampers situated in the boiler uptake. The by-pass contains a control unit which is an upward flow economiser section through which all the feed water to the boiler passes, this extracts excess heat from the gases as they pass through the by-pass section to the boiler uptake. When raising steam no feed water will be passing to the boiler, to prevent boiling off and possible damage occurring to the

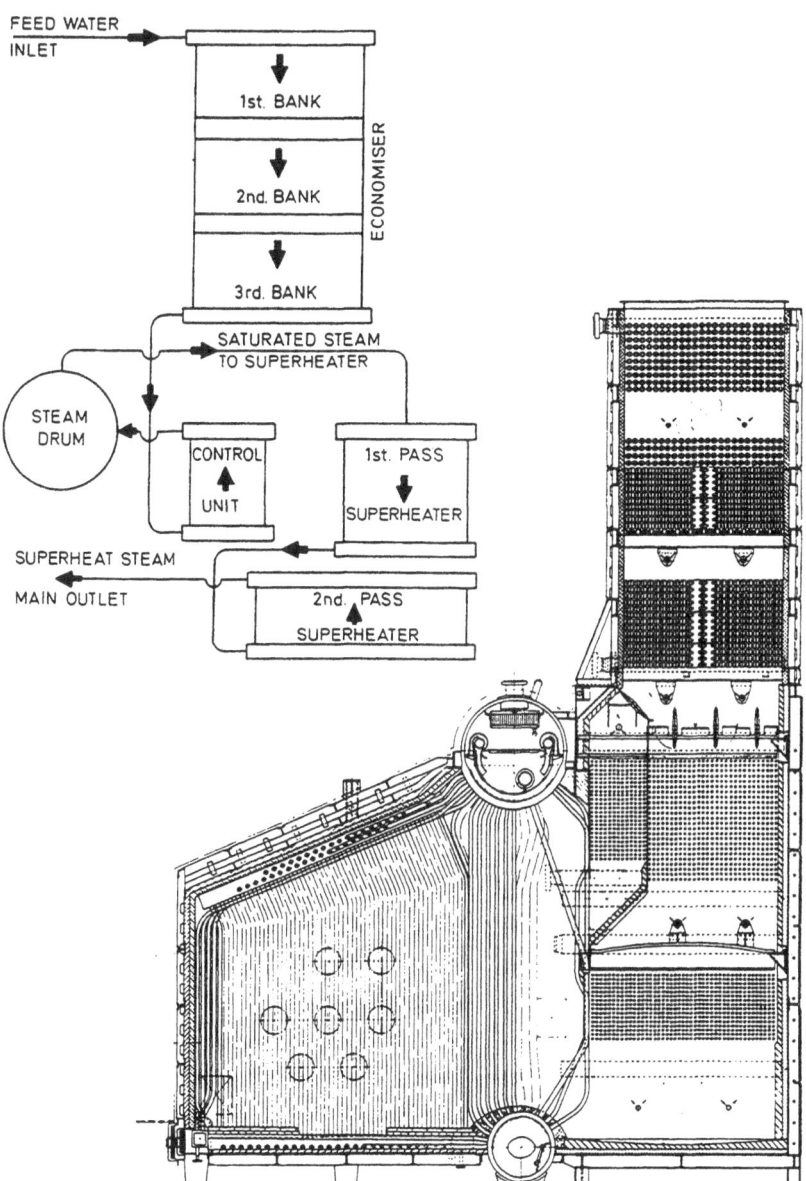

Fig. 1.9 FOSTER WHEELER ESD II BOILER AS FITTED TO QE2

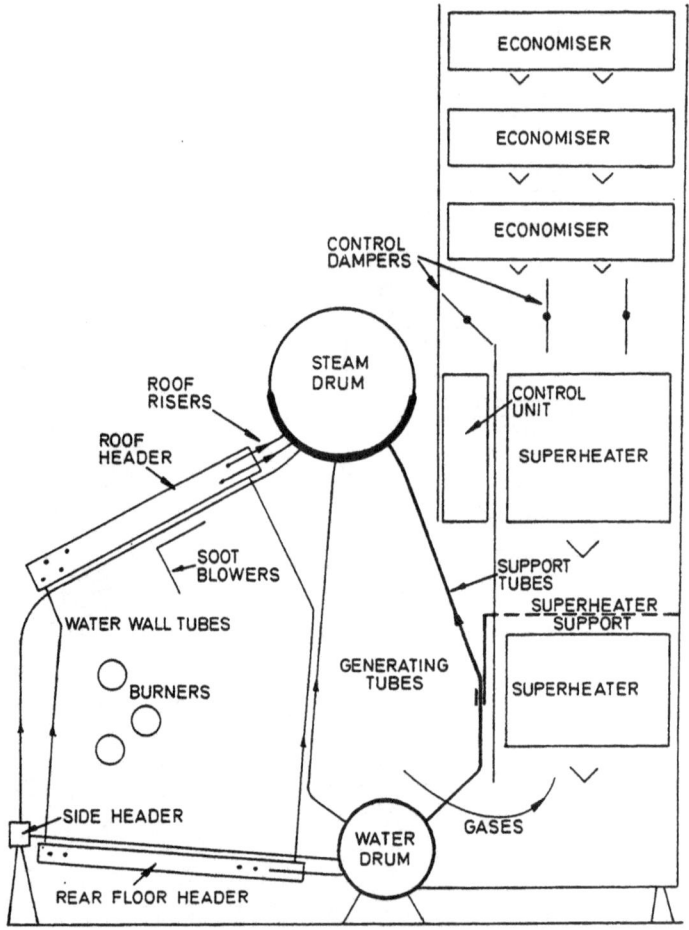

Fig. 1.10 ESD II BOILER

control unit, circulation through it is ensured by means of a balance leg connected to the water drum.

E.S.D. III Boiler

This type of boiler on its own is designed to supply all the steam requirements of a large tanker or container vessel developing con-

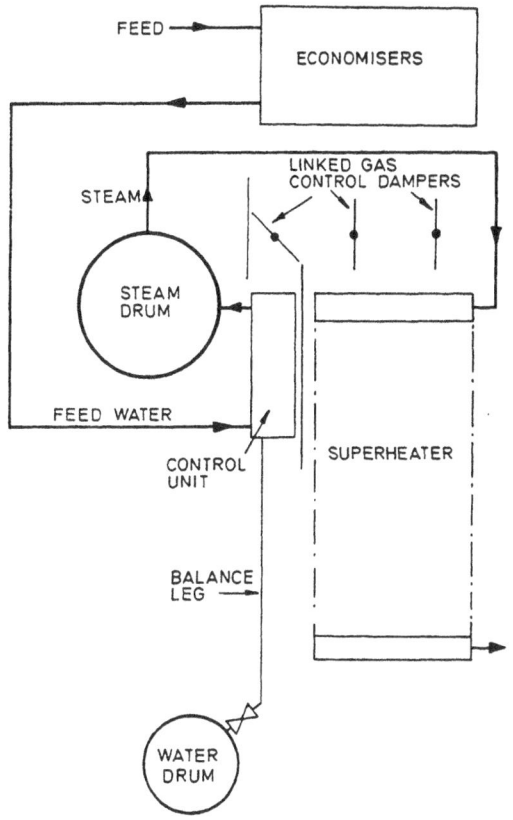

Fig. 1.11 STEAM TEMPERATURE CONTROL SYSTEM

siderable horse power. (A small auxiliary boiler may be fitted for stand-by purposes.)

Construction

The furnace is made up of close pitched water walls side, front and rear. Close pitching of these tubes gives good protection to the refractory fitted behind them. Floor tubes are protected by brick refractory. All tubes are expanded into drums and headers, and handhole plugs for tube inspection are provided in the headers.

Eight rows of screen tubes with large pitch at the bottom to give gas passage and ligament strength to drum form four rows of close pitched tubes from about the superheater bottom up. Behind the second row of screen tubes shims are provided to help keep the furnace gas tight.

Burners are arranged in the roof and the furnace gas passes down the furnace, through the openings formed by the screen tubes and then up over the superheaters.

The superheater, situated between the screen tubes and a side water wall, which again is made up of close pitched tubes that are expanded into the drums, is in two passes. The first pass consists of the two upper sections through which all the steam passes before going direct to the second pass, or through an attemperator coil in the steam drum water space, then to the second pass. In this way steam temperature control can be achieved.

To ensure positive circulation of water, large bore uncooled water wall and screen tube feeders are fitted external to the boiler.

A double casing (Fig. 1.12) is also fitted to the boiler which has the advantages of, (1) Reducing quantity of refractory required, (2) Serving as an air passage, (3) Making the furnace effectively gas tight. Its main disadvantages would be increasing boiler size and possibly making refractory repairs more inconvenient to carry out.

A typical set of figures for the E.S.D. III boiler are:

Evaporation	
normal	81 000 kg/h
maximum	100 000 kg/h
Pressure at superheater outlet	63 bar
Steam temperature	515° C
Feed temperature at economiser inlet	138° C
Air temperature, airheater outlet	120° C
CO_2 in flue gas	15%
Boiler efficiency	88.8%

Membrane Wall Boilers

This arrangement dispenses with most of the refractory in some boilers by the use of membrane walls. These consist of water tubes with a strip of steel welded between them to form a panel which is

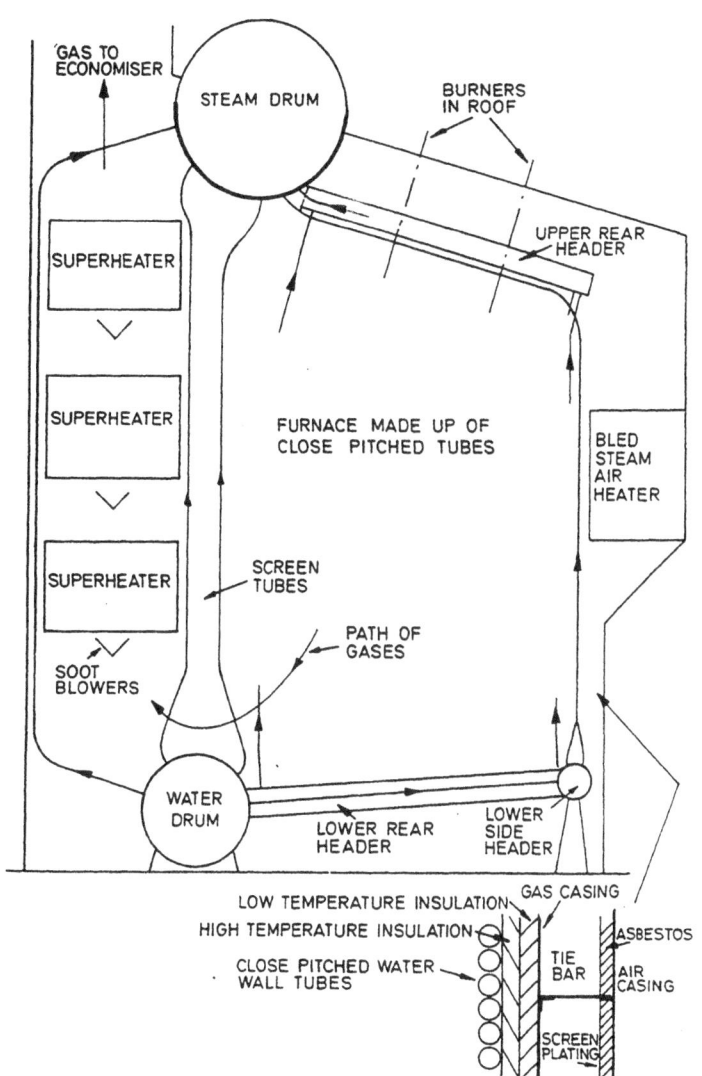

Fig. 1.12 ESD III BOILER

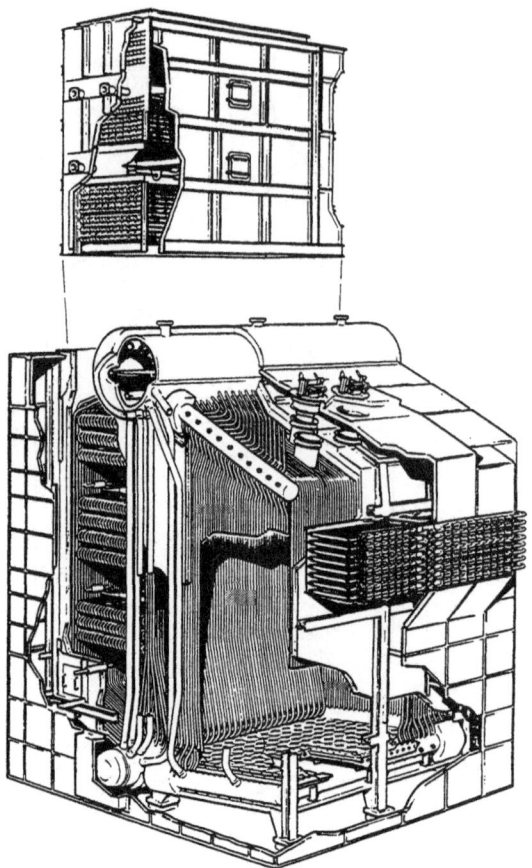

Fig. 1.13 FOSTER WHEELER ESD III MARINE BOILER

then bent into its required shape for the boiler. Alternatively the tubes may be finned, welding together the fins will produce the membrane construction. This is usually done by resistance welding the strips to the tube to produce a finned tube, then the finned tubes are placed in a jig, and an auto-welder joins adjacent fin tips. There is a great saving in cost with membrane tube walls since:

1.   Refractory is not required, there would be a saving in maintenance and first cost due to this. Also a saving in weight.

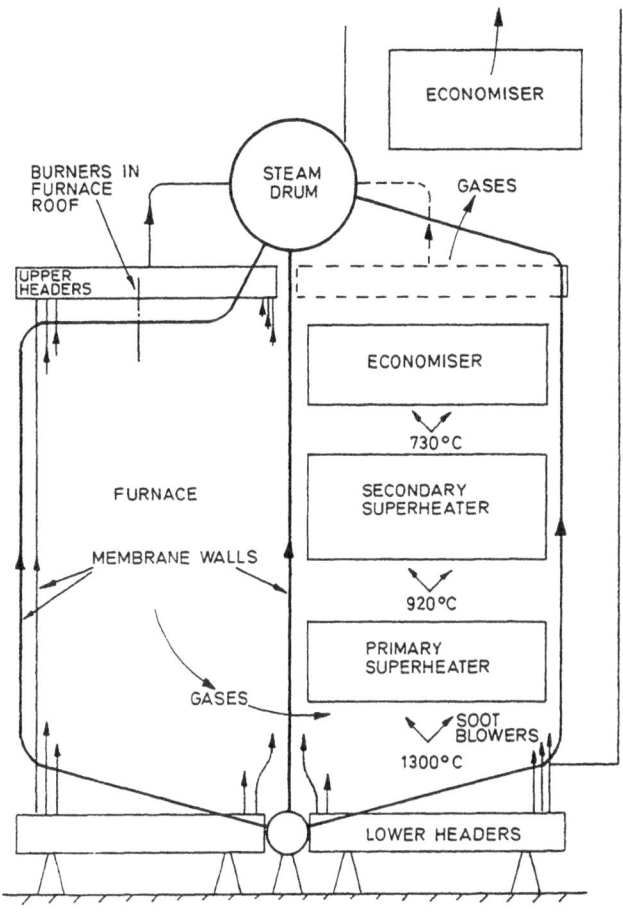

Fig. 1.14 MARINE RADIANT BOILER

2. Inner and outer casings are not required, furnace is effectively gas tight.

Fig. 1.14 is a diagrammatic arrangement of a radiant type boiler that uses membrane walls in its construction. The furnace is completely water cooled by the membrane tube walls which are usually welded to the drums and headers, thus eliminating tube expanding.

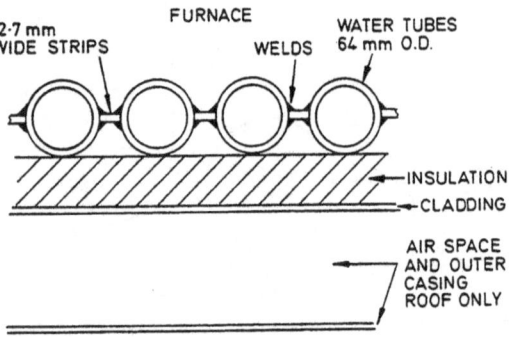

Fig. 1.15 MEMBRANE WALL

There is no conventional generating section, this is usual in two drum boilers, the gases pass instead through an opening at the lower end of the division wall then up over superheater and economiser surfaces to a gas air preheater (not shown) in the uptake.

External unheated large bore downcomers serve to feed the lower drums for the membranes and for steam temperature control, a coil attemperator may be situated in the water space of the drum, the attemperator would be connected between the primary and secondary superheater sections.

Burners are arranged in the roof.

The advantages of this type of boiler over the conventional are:

1. Increased efficiency.
2. Improved combustion.
3. Less excess air required.
4. Increased use of radiant heat, flame length increased by about 80%.
5. Refractory limited to burner quarls and other small items.
6. Gas tightness for furnace virtually ensured.
7. Inferior fuels can be burnt without them causing troublesome slagging of the refractory.

Note: The tube surface temperature and that of the steel connecting strip on the furnace facing side of the membrane walls is usually a maximum of about 340° C, hence no refractory to protect the membrane is required.

A typical Marine Radiant boiler of the Babcock type especially suited to high pressure steam cycles using membrane tube panel construction would be rated at:

Evaporation	45 500 kg/h to 226 000 kg/h
Steam pressure	48.3 bar to 110 bar
Steam temperature	470° C to 538° C
Excess air	5%
Efficiency	89% to 90.7%

With some economisers they can be an integral part of the boiler to such an extent whereby they exist to carry out a portion of the function normally accomplished by the generating tubes. Hence the term *bare tube steaming economiser* may be applied since up to about 5% of the feed water is evaporated in the economiser under normal steaming conditions.

Some economisers may have only a supply header to which the inlet ends of the bare tubes are welded, the outlets being connected to the boiler drum. This simplifies construction.

Gas flow is parallel to steam flow in the superheater section, hence hottest steam corresponds in location to the coolest gas. This provides for low metal temperature in service.

Roof firing with the minimum possible number of burners, this simplifies construction, operation and control, leads to increased residence time for the flame, better use, therefore, of the radiant heat, improved combustion and minimum excess air, 5% or below.

Reduced excess air means that the dew point of the vapour is reduced, less acid will be formed and corrosion is reduced.

If tube failure occurs two possible repairs can be carried out:

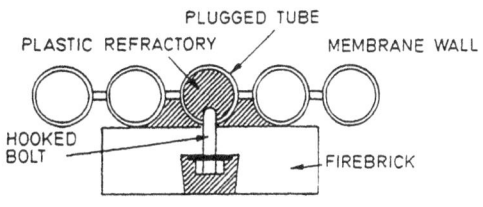

Fig. 1.16 MEMBRANE WALL REPAIR

1. Holes can be drilled at 300 mm intervals along the tube on the fireside for hooked firebrick retaining bolts as shown in Fig. 1.16. Sealing of the wall is done by plastic refractory clamped between the firebricks and damaged tube and packed into the damaged tube.

The plugged tube, protected in this way, should be returned to its original condition as soon as possible—reduced steaming may be necessary to ensure no further damage.

2. The defective portion of tube would be cut out by first drilling pilot holes and then sawing along the membrane and cutting across a diametral plane through the tube. The cut ends of the tube and membrane strip edges would be chamfered and a replacement tube, with half width membrane strip already welded on, would be welded in place. The boiler should then be hydraulically tested.

Superheater headers with stub tubes shop welded to them are shown in Fig. 1.17. They are externally arranged and will be protected from hot gases and flames. If damage were to occur it would be in the gas passage region or inside the tubes.

Tubes connected to the headers (superheater and economiser) may become damaged and have to be plugged, they are cut as indicated in the diagram so that a straight portion of tube remains for the fitting of a external plug—detail of which is shown.

For the control of final steam temperature of the E.S.D. III boiler and Membrane wall boilers of the radiant types refer to the control chapter.

Tangentially fired boiler

The Combustion Engineering (V2M9) boiler shown in Figs. 1.18 and 1.19 contains several new features, the most outstanding being the combustion arrangement. A burner is located at each of the four corners of the furnace and they are aligned to be tangential to a circle at the centre of the furnace. This imparts a rotary motion to the combustion gases giving improved turbulence and air-fuel mixing. This spiral effect keeps fuel particles in the furnace longer, so that by the time the hot gases reach the tubes all the fuel particles have been burned.

Because air-fuel mixing is so good, with normal operation employing steam atomising burners it is possible to operate with, it is claimed 1% excess air. This means reduced slagging and corrosion

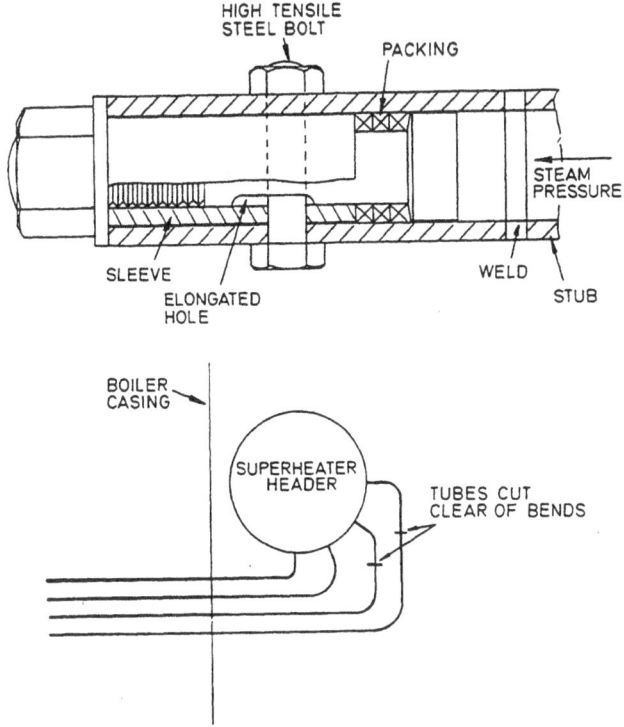

Fig. 1.17 TUBE PLUG

with higher efficiency.

The boiler has a unique support arrangement, being supported under the water drum and intermediate header of the side wall. By locating these supports at approximately mid-height, expansion of the upper part is up and of the lower part down, hence there is less relative expansion than in the bottom supported types.

Reheating

The reheat principle means that steam after expansion through the h.p. turbine is returned to the boiler for reheating to the initial superheat temperature. After, it is expanded through the i.p. and l.p. turbines. This arrangement can improve thermal efficiency by 4% or

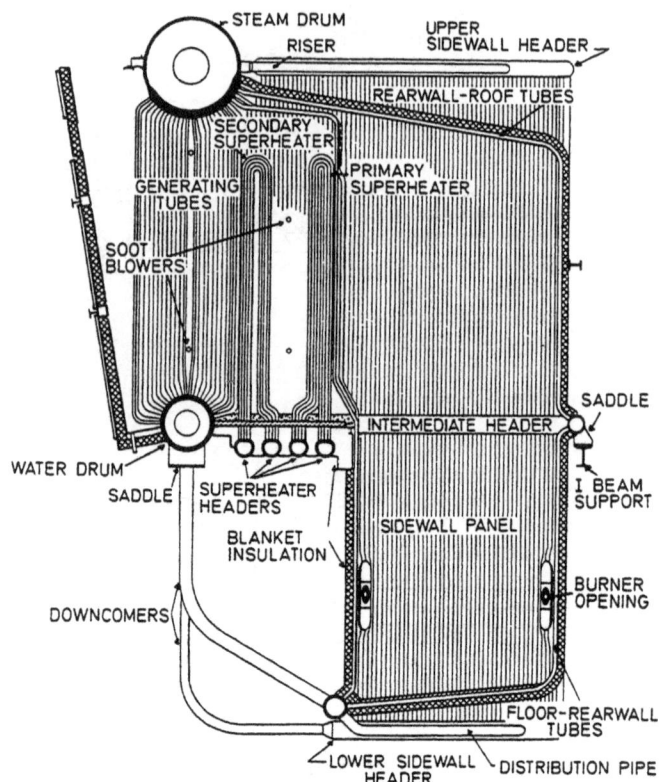

Fig. 1.18 TANGENTIAL FIRED BOILER (COMBUSTION ENGINEERING INC.)

more. It is likely, in the future, due to soaring fuel costs that reheating systems as a means of economy will increase in number.

Reheat pressure (*i.e.* h.p. turbine exhaust pressure) governs to a large extent the overall efficiency. Practical-theoretical considerations relating to this pressure are of paramount importance, *e.g.* boiler pipework, heating surface, bled steam, position of reheater in boiler (or separate reheat boiler), manoeuvring considerations, etc.

By increasing steam conditions and using reheat the steam flow rate can be reduced for the same power output.

A current type of reheat plant using a Babcock Marine Radiant Reheat boiler has reheat at 100 bar, 510° C which compared to a

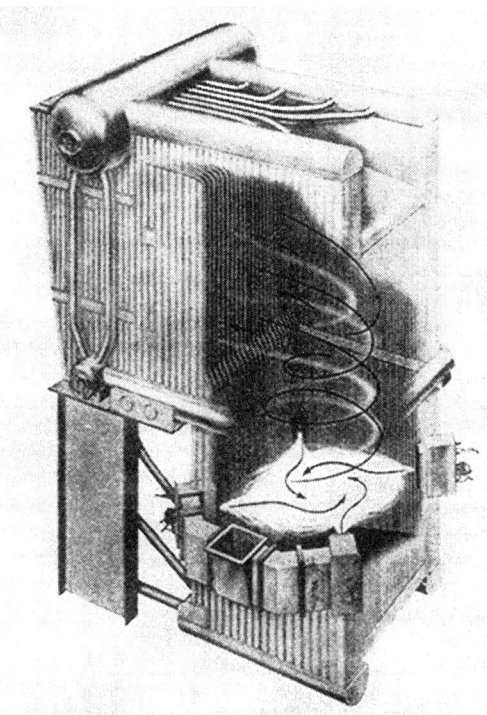

Fig. 1.19 TANGENTIAL FIRED BOILER (COMBUSTION ENGINEERING INC.)

straight cycle at 63 bar, 510° C gives a fuel saving of 7 to 8%. Approximate values of pressure and temperature are shown in Fig. 1.20, which is a diagrammatic arrangement of this type of boiler.

The roof fired boiler has a divided gas passage with gas flow controlled by dampers. Main steam temperature control is the same as the non-reheat radiant boiler but reheat temperature control is achieved by regulating the gas flow over the reheater. Damper position is controlled by a reheater outlet steam temperature control loop.

Reheater protection for no steam flow condition (manoeuvring mode) is accomplished by closing the dampers. Gas leakage past the closed dampers could cause overheating of the reheater but superheat stages in this part of the gas uptake cool the leakage gas sufficiently to preclude this.

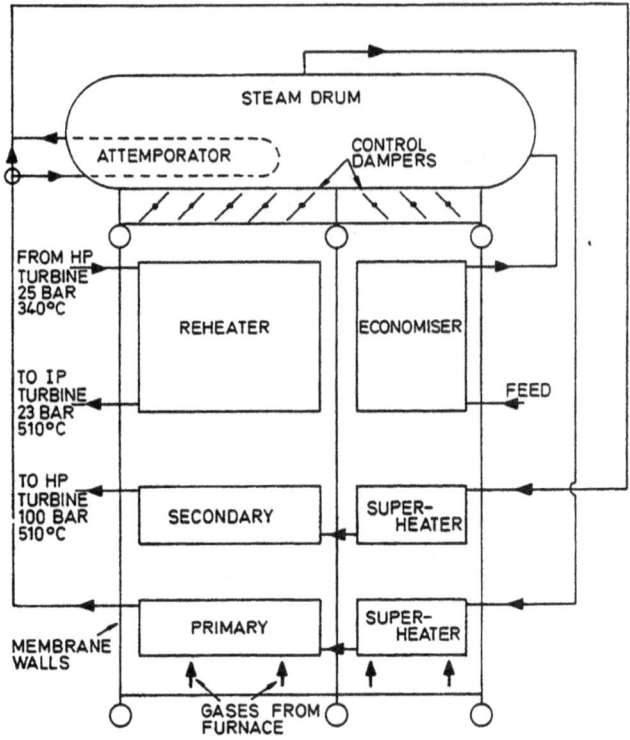

Fig. 1.20 LINE DIAGRAM OF MARINE RADIANT REHEAT BOILER

It is not possible to cover all the various types of water tube boiler in present use but the student should be aware of the existence of some of the other types not covered in any detail in this chapter. Fig. 1.21 is an attempt to help the student become aware of these other types.

1.   Yarrow five drum boiler with single furnace, gas flow over saturated and superheat sections controlled by dampers.

2.   Selectable superheat boiler with single furnace, the dotted line shows the path of the gases through the saturated section behind the superheater section, twin gas dampers are fitted to control the gas flow over these boiler sections which are separated by a vertical screen.

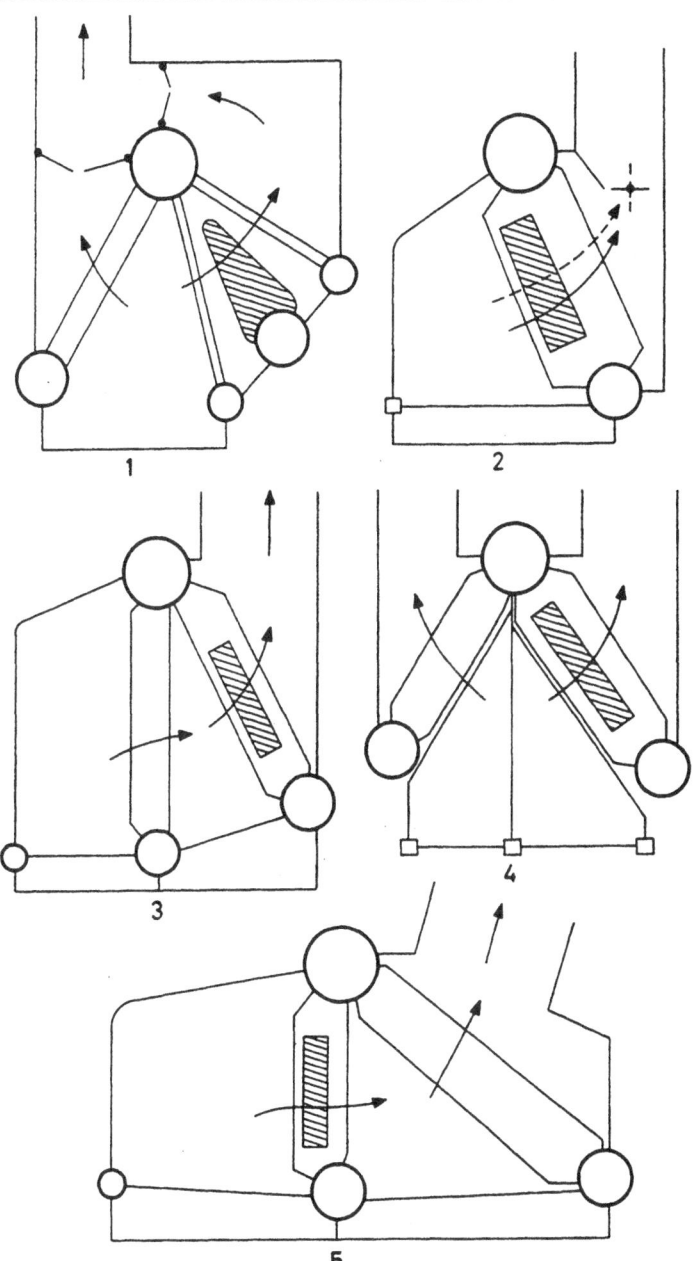

Fig. 1.21 VARIOUS TYPES OF WATER TUBE BOILERS

3. Foster Wheeler twin furnaces with a single uptake.
4. Babcock controlled superheat boiler with twin furnaces and a double uptake.
5. Babcock controlled superheat boiler with twin furnaces and a single gas uptake.

All those boilers fitted with twin furnaces control the final steam temperatures by using different firing rates in the individual furnaces.

CONTROL OF SUPERHEAT TEMPERATURE IN BOILERS AND DESUPERHEATING

Scotch Boilers

The production of superheated steam in this type of boiler is generally achieved by the use of small bore steam carrying element U tubes inserted into the gas carrying tubes. The element tubes are connected to supply and return headers fitted at the front of the boiler.

When raising steam the headers are first drained and the return header drain is left open to ensure circulation in the element U tubes, the supply header drain being closed.

Temperature control is by means of a mixing valve. The bulk of the steam passes through the superheater section, thus ensuring no overheating of the elements, and some steam passes straight from the boiler to the return line from the superheater header to mix with (and hence reduce the temperature) the superheated steam.

Water tube boilers

The methods of steam temperature control for water tube boilers are:

(1)   Damper control
Furnace gas flow across superheater sections in the boiler is controlled by dampers, the E.S.D. II, Babcock selectable superheat and other boilers use this method.

(2)   Differential firing
Two furnaces separated by a section of operating tubes or membrane wall, one with a superheater section the other without. The gas passage could be dual, merging into one, or single.

(3)   A combination of (1) and (2) could be employed.

In addition to the foregoing methods attemperation is often used, the attemperator may be situated in the air passageways—hence air attemperator, or in the water space of a boiler drum. An attemperator removes some of the superheat in the steam but it is not to be confused with a desuperheater.

Desuperheaters remove the superheat in the steam that is to be used for auxiliary purposes, the reason is that if high temperature steam was used for auxiliaries then the materials used would have to be capable of withstanding high temperatures, this leads to increased first cost. Desuperheaters are generally coils situated in the water spaces that are supplied with steam from the superheater section outlet, this type of desuperheater is generally called internal. If this type of desuperheater develops a leakage then water loss from the boiler will occur since the boiler pressure is greater than the steam pressure in the desuperheater coil. This water loss could result in water hammer and subsequent damage in the auxiliary system.

External desuperheaters may be of the spray type. Fig. 1.22 shows the arrangement, in which feed water of low solids content at 3.5 bar above steam pressure, is supplied at a controlled rate into a vertically arranged vessel to mix with superheated steam, the excess water which is kept to a minimum, collects at the bottom of the vessel and passes out through the drain and trap, the desuperheated steam moves up the annular space formed by the mixing compartment and the outer container to the auxiliary steam supply line.

Another type of external desuperheater is shown in Fig. 1.22, this is really part of the boiler externally arranged. It has a steam and water connection to the boiler drum so that circulation through the desuperheater takes place, taking heat from the superheated steam in the solid drawn steel U shaped tubes. A blow down valve is fitted in order to remove any sludge that may accumulate in the lower portion of the vessel. This type of desuperheater is for large steam demands since the three pass U tube arrangement would be too large to accommodate in the water space of the steam drum.

For extremely large demands for saturated steam an auxiliary boiler would probably be the best arrangement or one whereby the superheater can be completely by-passed without causing any damage by steam starvation.

Often due to incorrect boiler operation, components malfunctioning, or when raising steam, high steam temperatures at outlet from the boiler can be realised.

High steam temperatures mean high metal temperatures and this leads to a reduced life-span for pressure components due to creep. (If extra high metal temperatures are encountered corrosion due to Sodium and Vanadium bearing ash can also be a problem.) Hence high steam temperatures should be avoided.

A *rapid* rise in steam temperature could be caused by a fire breaking out in the superheater unit due to accumulation of sooty deposits. Other causes of high temperature could be:

(a) Partial or slightly reduced steam flow through the superheater, this could be caused by (i) taking steam for auxiliaries direct from the boiler drum, (ii) leaking attemperator in the air space.

(b) Increased gas temperature due to the burning of extra fuel, this could be caused by (i) steam demand increasing, (ii) feed heater out of service, (iii) steam air heater out of service, (iv) by-passing a gas airheater, (v) dirty gas airheater, (vi) high delivery rate from fan, (iii), (iv) and (v) mean no preheated air for the furnace, this can have a cooling effect which has to be counteracted by burning more fuel which means more excess air as in (vi) which in turn leads to a greater volume of hot gas passing through the boiler per unit time.

Naturally the position of the superheater section affects the result. If the superheater is partially screened by generating tubes then it will be affected by radiant heat as well as gas temperature, hence if more fuel is being burned a greater quantity of radiant heat is available, this would not affect a superheater situated in the gas uptakes of the boiler remote from the furnace.

Low steam temperature can also be encountered. If there is a *rapid* fall in steam temperature this would most likely be due to (1) Boiler priming, (2) Sudden leakage developing in the attemperator in the water space.

More gradual falls in steam temperature could be due to (a) accumulation of deposits on the superheater leading to poor heat transfer from gas to steam, (b) accumulation of deposits inside the superheater tubes due to carrying too high a water level, defective dry pipe, inadequate baffling, oil in the water or relatively high boiler water density, (c) reduction in steam demand.

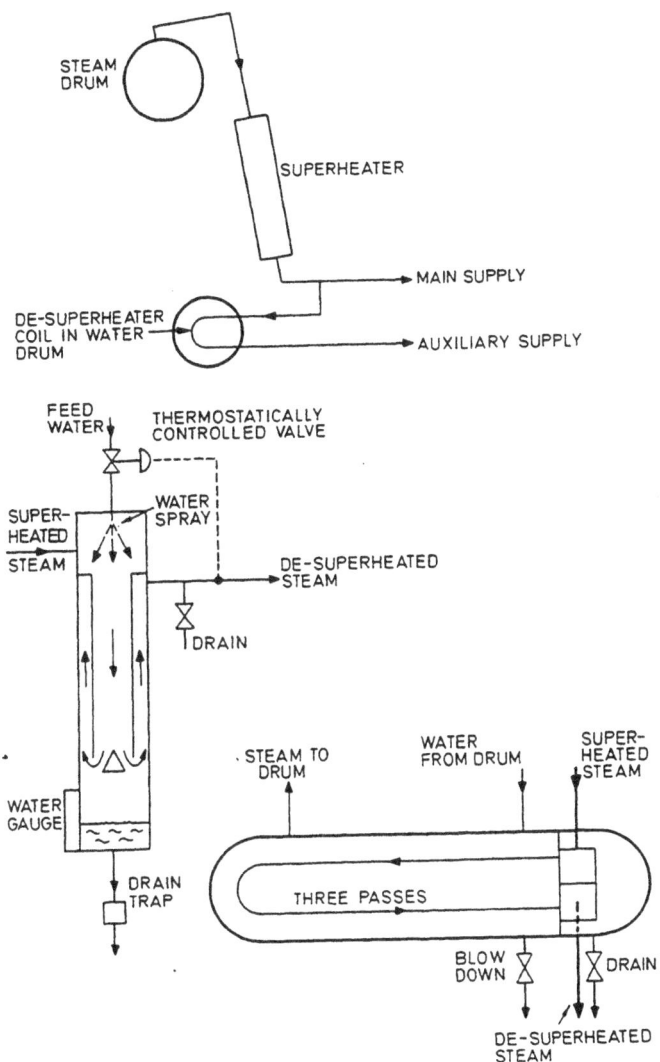

Fig. 1.22 EXTERNAL DE–SUPERHEATER SURFACE TYPE

REFRACTORY

What it is

A refractory is a material in solid form that is capable of maintaining its shape at high temperatures. Refractoriness may be defined from the foregoing as ability to maintain shape at elevated temperatures.

Various types of refractories are used in modern marine boilers they could be classified simply as follows:

Firebrick

Usually these are made from naturally occurring clays which contain alumina and silica, the clay is shaped into brick form and then fired in a kiln. The refractoriness and resistance to fuel ash increases with the increase of alumina content.

Plastic refractory

Small amounts of plastic clay can be added to calcined, (*i.e.* roasted until crumbly) fireclay which has been crushed and graded, the amount varies up to 20%. This type of refractory can be moulded into position—hence mouldable. (Note: mouldable refractories have to be fired in the furnace, castable refractories are air drying like structural concrete.)

Plastic chrome ore is a refractory that is used considerably for monolithic linings (*i.e.* one piece) in modern water tube boilers.

Insulating material

Insulating blocks, bricks, sheets and powder are usually second line refractory, *i.e.* they are behind the furnace refractory which is exposed to the burner flame. The material could be asbestos millboard, magnesia asbestos, calcined magnesia blocks or diatomite blocks, diatomite is porous and siliceous, it may be used as a powder, bonded with clay and made into bricks.

What it does

The refractory materials used in marine boilers perform the following functions:

1.   Gives protection to the furnace casing.
2.   Reduces heat loss from the boiler.

3. Gives protection to furnace tubes.
4. Acts as a reservoir of heat.

Defects

Refractory exposed to the burner flame is subject to spalling. Spalling is simply loss of refractory material from the hot surface. This could be caused by (1) Fluctuations in temperature resulting in cracking, (2) Mechanical straining due to furnace pressure variation (panting of the boiler is an example, panting could be caused by water in the oil, burners malfunctioning, incorrect oil pressure and temperature), (3) Defective construction, (4) Slag attack, this is a chemical reaction between the ash from the fuel and the refractory. Sodium and Vanadium constituents in the ash could be mainly responsible, they produce a reactive fluid slag which rapidly erodes the furnace lining. If the slag which is formed is viscous it may just adhere to the refractory. Sea water in the fuel can lead to slag attack since it contains sodium and calcium salts.

To avoid crushing of the refractory due to expansion, allowances have to be made, also to prevent undue loading due to its own weight. Adequate support has to be given. Fig. 1.23 shows arrangements of refractory. Fig. 1.23 (a) is a section through a fully studded water wall that is to be found in some Babcock and Wilcox boilers, the tubes have steel studs resistance welded to them in order to retain and give support to the plastic chrome ore refractory. Fig. 1.23 (b) shows a bare wall arrangement of firebricks with supporting bolt. Fig. 1.23 (c) gives the cover arrangement for a front and side wall showing the allowance for expansion and protection.

STEAM TO STEAM GENERATION

In vessels which are fitted with water tube boilers a protection system of steam to steam generation may be used instead of desuperheaters and reducing valves, etc.

The generator is really a boiler made of carbon steel in which are steam carrying heating coils or U tubes made of carbon steel or alloy steel. It produces from feed water (that may well be untreated fresh water) steam at low pressure for auxiliary purposes. Pressures may range from 4.5 to 18.5 bar and output may be from 450 to 45 000 kg/h depending upon generator size.

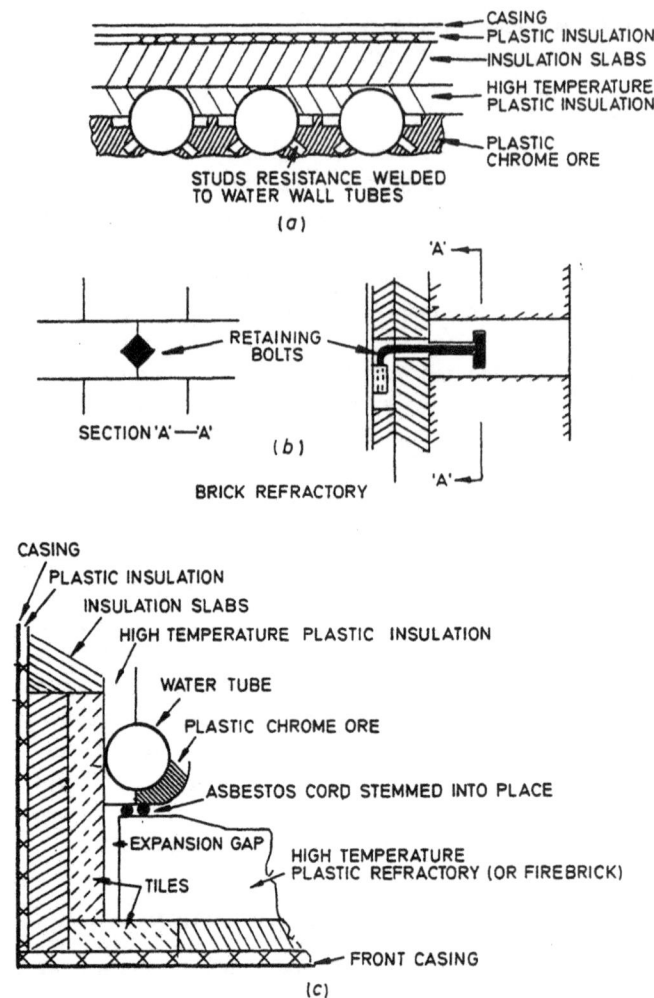

CASING
PLASTIC INSULATION
INSULATION SLABS
HIGH TEMPERATURE
PLASTIC INSULATION
PLASTIC
CHROME ORE

STUDS RESISTANCE WELDED
TO WATER WALL TUBES

(a)

RETAINING
BOLTS

SECTION 'A'—'A'   (b)

BRICK REFRACTORY

CASING
PLASTIC INSULATION
INSULATION SLABS
HIGH TEMPERATURE PLASTIC INSULATION
WATER TUBE
PLASTIC CHROME ORE
ASBESTOS CORD STEMMED INTO PLACE
EXPANSION GAP
HIGH TEMPERATURE
PLASTIC REFRACTORY (OR FIREBRICK)
TILES
FRONT CASING

(c)

Fig. 1.23 BOILER REFRACTORY

The main advantages of the system are (1) Untreated water can be used as feed. (2) Auxiliary system contaminants do not enter the main boilers, (3) Increased economy since (a) if steam from water tube boilers was used for auxiliaries, make up feed would have to be evaporated fresh or salt water and (b) reduced main boiler maintenance.

Fig. 1.24 shows such an arrangement with automatic control and feed heating to provide an economy. The generator is fitted with the usual boiler mountings such as two safety valves, gauge glass, feed regulator, blow down, scum valve, salinity test cock and chemical injection point, etc. In order to minimise risk of carry over an internal dry pipe and baffle arrangement is provided.

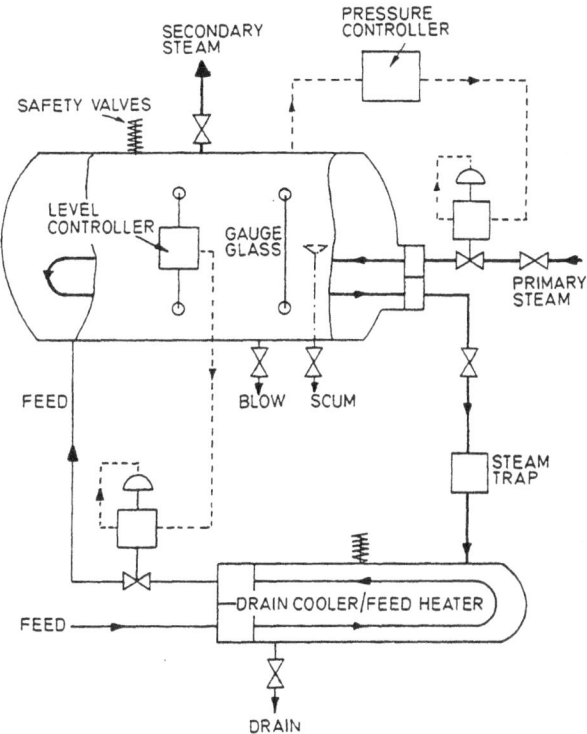

Fig. 1.24 STEAM TO STEAM GENERATOR

Double evaporation boilers

These are another version of steam to steam generation.

The primary system is completely closed and takes the form of a water tube boiler consisting of steam and water drums connected by tubes. The secondary system consists of a steam generator drum which produces steam at low pressure, superheated if desired, in relatively large quantities for auxiliary purposes. Both systems are welded throughout, with tubes expanded into position. The arrangement provides maximum security against problems caused by impurities such as salts and oil in the boiler water.

Data:

Primary system
Working pressure 50 bar
Water content 4.7 to 11.5 tonnes.

Secondary system
Working pressure 16.5 to 18 bar
Evaporation rate 15 000 to 30 000 kg/h
Superheated steam 230° C (if fitted).

Although mention was made that the primary system is completely closed, obviously some provision is made for make up in the event of leakage.

COMBUSTION EQUIPMENT

Good combustion is essential for the efficient running of the boiler it gives the best possible heat release and the minimum amount of deposits upon the heating surfaces. To ascertain if the combustion is good we measure the % CO_2 content and observe the appearance of the gases.

If the % CO_2 content is high as practicable and the gases are in a non smoky condition then the combustion of the fuel is correct. With a high % CO_2 content the % excess air required for combustion will be low and this results in improved boiler efficiency since less heat is taken from the burning fuel by the small amount of excess air. If the excess air supply is increased then the % CO_2 content of the gases will fall. The curve shown in Fig. 1.26 shows the general trend.

In modern plants carbon dioxide and oxygen content of the flue gases is measured. Relationship between O_2 content of flue gases and excess air is nearly independent of the type of fuel used in the boiler. Hence O_2 content of the flue gases is being used more often in

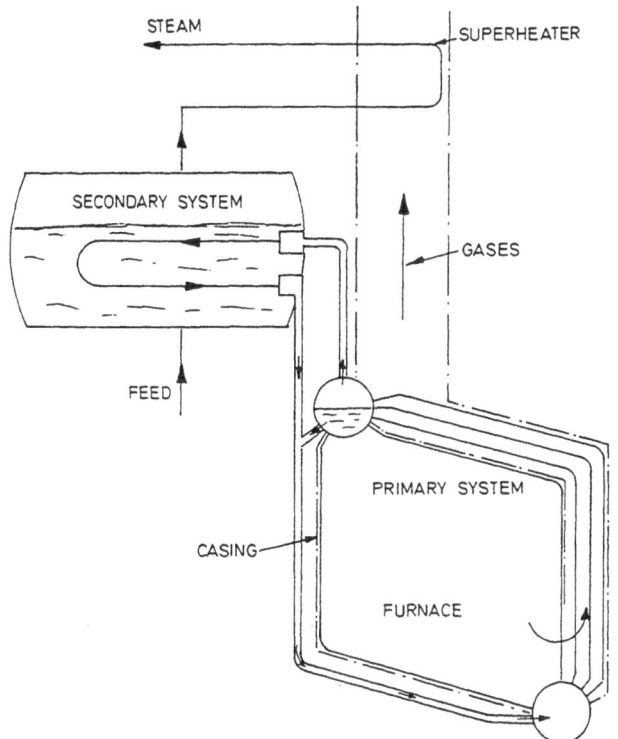

STEAM

SUPERHEATER

SECONDARY SYSTEM

GASES

FEED

PRIMARY SYSTEM

CASING

FURNACE

Fig. 1.25 DOUBLE EVAPORATION BOILER

modern plants to give indication of fuel burning efficiency.

Oxygen content is a better parameter to employ with automatic control of boilers since CO_2 content can vary depending upon fuel and combustion conditions.

Oil fuel atomisers
Simple pressure jet

Fig. 1.27 (a) shows diagrammatically the principle of the simple pressure jet atomiser. Oil fuel at a pressure of not less than 8 bar is supplied to the tangentially arranged ports, wherein it falls in

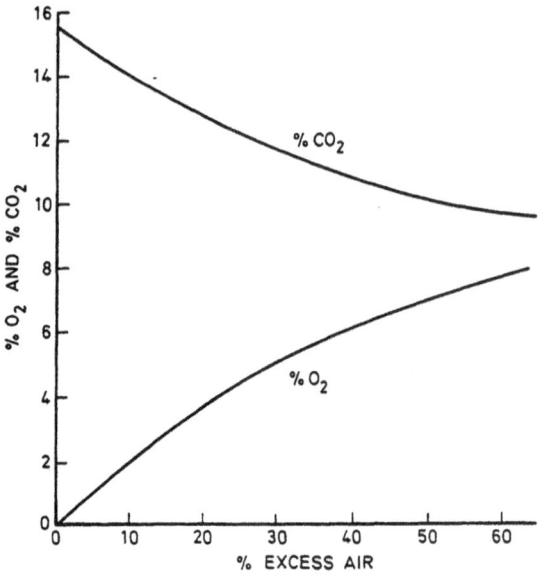

Fig. 1.26

•pressure, resulting in the oil swirling around at high velocity inside the chamber. The greater the pressure drop the greater would be the velocity, but generally supply oil pressure would not be greater than 25 bar—high pressures create pumping and sealing problems.

At outlet from the short sharp edged orifice (care must be taken not to damage the orifice during cleaning) a hollow expanding cone of fuel droplets is produced whose initial film thickness reduces as oil supply pressure increases.

The following features of the simple pressure jet atomiser should be noted:

1.   Simple, inexpensive and robust.
2.   No moving parts, hence no possibility of seizure.
3.   Large range of droplet size for one pressure.
4.   Turn down ratio is low, about 2.5 to 1. This is the ratio of burner throughput maximum to minimum. If say the maximum throughput is 2500 kg/h then with a turndown ratio of 2.5 to 1 the minimum throughput will be 1000 kg/h.

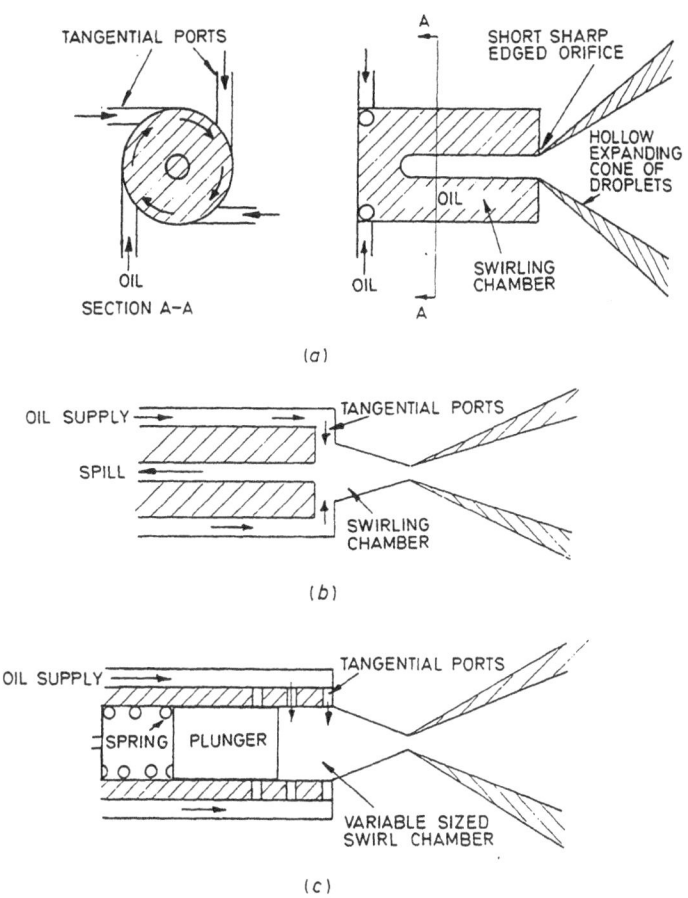

Fig. 1.27 OIL FUEL ATOMISERS

5. Maximum throughput about 3200 kg/h.
6. No alteration in air register arrangement for all outputs—this is a consequence of the low turndown ratio of this type of burner.
7. To vary throughput the pressure must be altered, this results in variation in droplet size, hence atomisation and combustion will be affected.

Spill type of pressure jet atomiser   Fig. 1.27 (b)

With this type of atomiser throughput is varied by adjusting the amount of 'spill,' *i.e.* the return flow from the burner.

Constant delivery pressure type

With this type, if the amount of spill is increased the throughput will be reduced, but this increases the pressure drop across the tangential ports. Hence atomisation is increased and combustion will be easier. For this reason the burner has a high turn down ratio, up to about 20 : 1 is claimed.

Constant differential pressure type

With the differential pressure between delivery and spill kept constant the pressure drop across the tangential ports is kept constant and the rotational velocity of the fuel in the swirl chamber will not alter. This will keep fuel cone angle and atomisation constant.

The main disadvantage with these types of burner is the large quantity of hot oil being returned, it may prove difficult to keep control over oil temperature.

Plunger type of pressure jet atomiser

Fig. 1.27 (c) shows the principle of operation of this type of burner, as the oil supply pressure is increased the spring loaded plunger moves to uncover extra tangential oil entry holes. The pressure drop and hence the rotational velocity of the oil remains nearly constant.

Rotary or spinning cup atomisers

Fig. 1.28 shows the arrangement in principle. It consists of a motor driven fan, metering pump and fuel cup.

The fuel cup which rotates at 70 to 100 rev/s is supplied with oil at low pressure (1.7 to 4.5 bar) from the metering pump which is used to control throughput. Due to centrifugal force and opposite swirl of air to fuel the oil film leaving the cup is rapidly broken down into relatively even sized droplets.

Main features are:

1.   High output possible, up to 3600 kg/h.
2.   Low oil supply pressure.

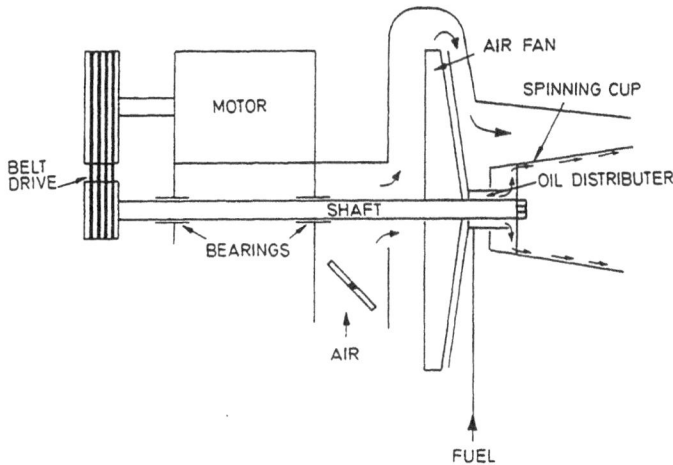

Fig. 1.28 SPINNING CUP ATOMISER

3. At low throughputs atomisation due to reduced oil film thickness is improved.

4. Wide turn down ratio, up to 20 : 1 possible.

5. Oil viscosity need only be reduced to 400s Redwood number 1 for satisfactory operation.

Condition of burners, oil condition pressure and temperature, condition of air registers, air supply pressure and temperature are all factors which can influence combustion.

Burners: If these are dirty or the sprayer plates are damaged then atomisation of the fuel will be affected.

Oil: If the oil is dirty it can foul up the burners. (Filters are provided in the oil supply lines to remove most of the dirt particles but filters can get damaged. Ideally the mesh in the last filter should be smaller than the holes in the burner sprayer plate.)

Water in the oil can affect combustion, it could lead to the burners being extinguished and a dangerous situation arising. It could also produce panting which can result in structural defects.

If the oil temperature is too low the oil does not readily atomise since its viscosity will be high, this could cause flame impingement, overheating, tube and refractory failure. If the oil temperature is too high the burner tip becomes too hot and excessive carbon deposits

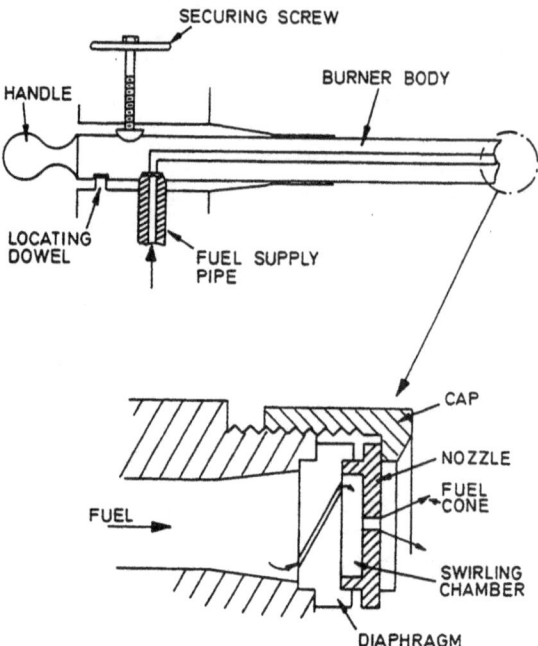

Fig. 1.29 OIL FUEL BURNER (WALLSEND–HOWDEN)

can then be formed on the tip causing spray defects, these could again lead to flame impingement on adjacent refractory and damage could also occur to the air swirlers.

Oil pressure is also important since it affects atomisation and lengths of spray jets.

Air registers: Good mixing of the fuel particles with the air is essential, hence the condition of the air registers and their swirling devices are important, if they are damaged mechanically or by corrosion then the air flow will be affected.

The flame stabiliser shown in Fig. 1.30 (a) achieves flame stability by producing eddy streams of air downstream. Fig. 1.30 (b) shows the air flow pattern and the air reversals, this will give what is often called a 'suspended flame,' 15 to 60 cm downstream of the swirler stabiliser. Flame stability is essential to ensure that the oil-air mixture does not have a higher velocity than the flame, the air reversals slow the combustible mixture down.

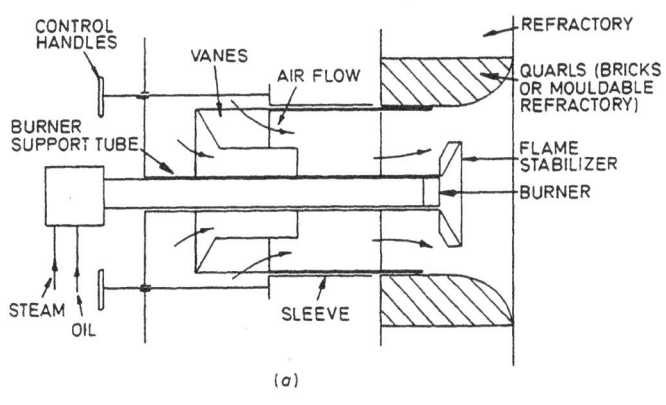

(a)

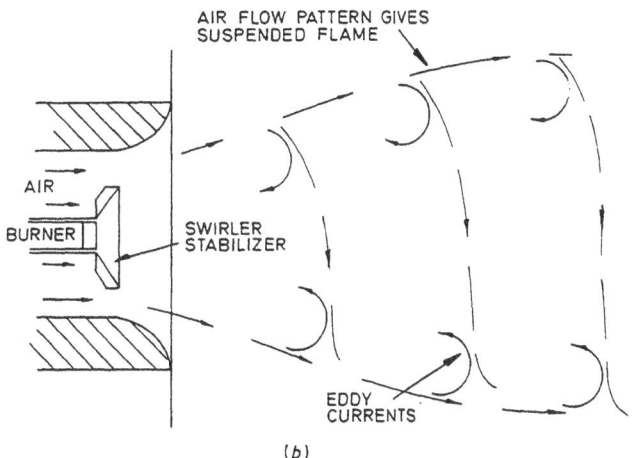

(b)

Fig. 1.30 AIR REGISTER FOR BOILER FRONT

Air: The excess air supply is governed mainly by the air pressure and if this is incorrect combustion will be incorrect.

Fig. 1.29 shows a simple burner arrangement for a boiler, preheated pressurised fuel is supplied to the burner tip which produces a cone of finely divided fuel particles that mix with the air supplied around the steel burner body into the furnace. A safety point of some importance is the oil fuel valve arrangement. It is *impossible* to remove the burner from the supporting tube unless the oil fuel is shut off, this greatly reducers risk of oil spillage in the region of the boiler front.

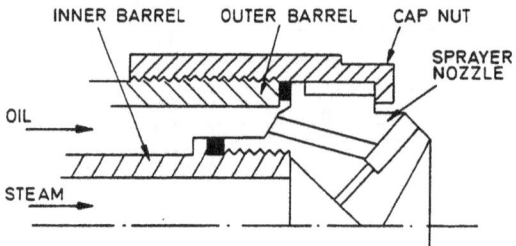

Fig. 1.31 'Y' JET TIP STEAM ATOMISING OIL BURNER

Fig. 1.31 shows the business end of the relatively recent Y-jet steam atomising oil burner that is finding increased favour for water tube boilers for the following reasons:

1. Deposits are greatly reduced, hence soot blowing and water washing of the gas surfaces need not be carried out as frequently as before (18 months or more between cleaning has been found possible).
2. Atomisation and combustion are greatly improved.
3. % CO_2 reading is increased (% O_2 reading has been lowered to 1% or below) hence boiler efficiency is greatly improved.
4. Atomisation is excellent over a wide range of loads and the turn-down ratio is as high as 20 : 1 and above.
5. With improved combustion, and turndown ratio, refractory problems are reduced.

The major disadvantage of this type of burner is that it uses steam—which means water and fuel—but the steam consumption in the latest type of steam atomiser is extremely small, less then 2% of the oil consumption at peak loads, and oil supply pressure as low as 1.3 bar can be used. In the event of steam supply failing the burner can be easily reverted to the pressure-jet principle. A steam control valve may be fitted to reduce the steam pressure at low loads.

An improved steam or air assisted atomiser incorporating acoustics is new to the field. A dense high frequency (14 kHz) vibrating wall of pressure waves is formed by directing the steam (or air) at high velocity through an annulus into a resonator chamber. The result is a vibrating cylindrical field through which the oil fuel must pass and in so doing is effectively atomised before igniting. Advantages claimed are (1) reduced maintenance, (2) wide turn down ratio, (3) excellent flame stability at all loads.

Boiler efficiency

Due to soaring fuel costs it has become extremely important that boiler efficiencies are maintained at the highest value possible under operating conditions, and that the designer produces a boiler arrangement such that it is simple, safe, highly efficient, capable of burning widely different oils and reliable.

Combustion equations, which are essential in boiler efficiency calculations, have been covered in detail in Volume 8, but some reiteration is necessary.

Combustion of Carbon

	$C + O_2$	$\rightarrow$	CO_2
Relative masses	$12 + (16 \times 2)$	$\rightarrow$	44
Relative masses	$1 + 2\frac{2}{3}$	$\rightarrow$	$3\frac{2}{3}$

Hence 1 kg of Carbon requires $2\frac{2}{3}$ kg of Oxygen and forms $3\frac{2}{3}$ kg of Carbon Dioxide liberating 33.7 MJ of heat in the process.

Combustion of Hydrogen

	$2H_2 + O_2$	$\rightarrow$	$2H_2O$
Relative masses	$(2 \times 1 \times 2) + (16 \times 2)$	$\rightarrow$	(2×18)
Relative masses	$1 + 8$	$\rightarrow$	9

Hence 1 kg of Hydrogen requires 8 kg of Oxygen and forms 9 kg of water vapour liberating 144.4 MJ of heat in the process.

Combustion of Sulphur

	$S + O_2$	$\rightarrow$	SO_2
Relative masses	$32 + (16 \times 2)$	$\rightarrow$	64
Relative masses	$1 + 1$	$\rightarrow$	2

Hence 1 kg of Sulphur requires 1 kg of Oxygen and forms 2 kg of Sulphur Dioxide liberating 9.32 MJ/kg of heat in the process.

Calorific value

Higher calorific value h.c.v. $= 33.7C + 144.4(H_2 - O_2/8) + 9.32S$

Lower calorific value l.c.v. $= h.c.v. - 2.465(H_2O)$

In practice the l.c.v. is chosen since it represents a more realistic value. $(H_2 - O_2/8)$ is the available Hydrogen since the fuel analysis will contain some water, and the Hydrogen in the analysis is the total. $2.465 (H_2O)$ is the heat taken from the burning fuel to turn all the H_2O in the fuel and that formed during combustion into steam.

Typical practical example

Fuel analysis: Carbon 88%, Hydrogen 10%,
       Sulphur 1.2%, Oxygen 0.4%
Fuel oil: h.c.v. 43 MJ/kg
Ambient air temperature: 38° C
Funnel gas outlet temperature: 116° C
Specific heat of funnel gas 1.06 kJ/kg° C
Excess air supplied 5%

Oxygen required for combustion of Carbon	$= 0.88 \times 2\frac{2}{3}$
	$= 2.35$ kg
Oxygen required for combustion of Hydrogen	$= 0.1 \times 8$
	$= 0.8$ kg
Oxygen required for combustion of Sulphur	$= 0.012 \times 1$
	$= 0.012$ kg

Total oxygen required $= 2.35 + 0.8 + 0.012 - 0.004$
              $= 3.158$

Theoretical air required	$= 3.158 \times 100/23$
(n.b. air contains	$= 13.7$ kg/kg of fuel
23% Oxygen)	
Actual air supplied	$= 1.05 \times 13.7$
	$= 14.39$ kg/kg of fuel
Amount of gas/kg of fuel	$= 15.39$ kg
Heat carried away by gases	$= 15.39 \times 1.06 (116 - 38)$
	$= 1270$ kJ/kg, *i.e.* 1.27 MJ/kg

Moisture in funnel gases $= 0.1 \times 9$
                                    $= 0.9$ kg/kg of fuel
Heat carried away by moisture $= 0.9 \times 2.465$
                                    $= 2.22$ MJ/kg

Heat balance

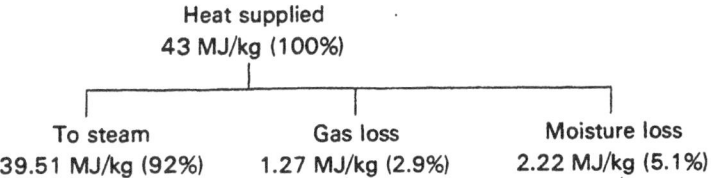

Heat supplied
43 MJ/kg (100%)

To steam	Gas loss	Moisture loss
39.51 MJ/kg (92%)	1.27 MJ/kg (2.9%)	2.22 MJ/kg (5.1%)

Hence boiler efficiency is 92% (39.51/43 x 100%)
A more accurate value could be 91% since some loss to radiation would occur.

BOILER MOUNTINGS

1. Blow down valve

Fig. 1.32 is a boiler blow down of the parallel spring loaded slide valve type suitable for high pressure operation. The double seating arrangement reduces risk of leakage and as an additional precaution the valves may be fitted in pairs.

A spanner fits over the square end of the pinion spindle *only* when the valve is in the closed position, when the valve is opened the spanner cannot be removed from the spindle. This serves as a safeguard, for if a number of boilers were connected to the one blow down line it is impossible to have more than one blow down valve open, since there is only one spanner for all the valves, hence boiler contents could not be blown into an empty boiler.

Since the valves and seats are going to be subjected to corrosive and erosive conditions, the material used must be non-corrosive and hard. Also the valve body is subjected to high pressure hence cast steel or equivalent material would be required.

2. Feed water regulators

These are essential for boilers with high evaporative rates such as water tube boilers. The advantages of feed water regulation are:

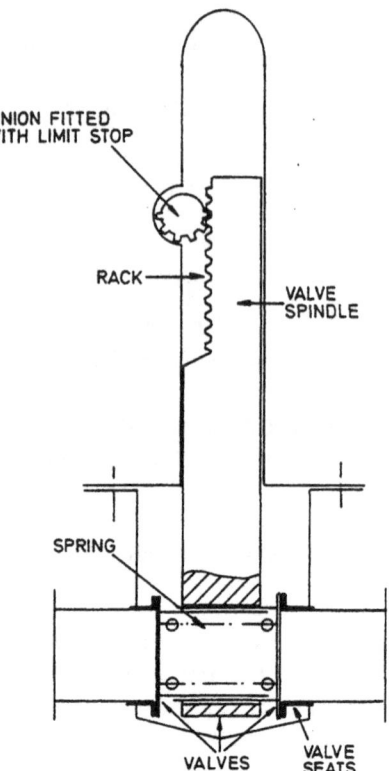

Fig. 1.32 BOILER BLOW DOWN VALVE

(a)   Reduced risk of priming, hence this safeguards the superheater.
(b)   Drier steam, hence the solids content of the steam will be reduced and contamination of steam piping is reduced.
(c)   Reduced risk of water shortage, overheating and failure.
(d)   Maintains steadier steaming conditions.
(e)   Relieves personnel of the arduous duty of feed water regulation.

The Weir robot feed controller shown in Fig. 1.33 gives sensitive control over the water level and maintains it between two extremes, no load condition and maximum load. It responds rapidly to changes in water level.

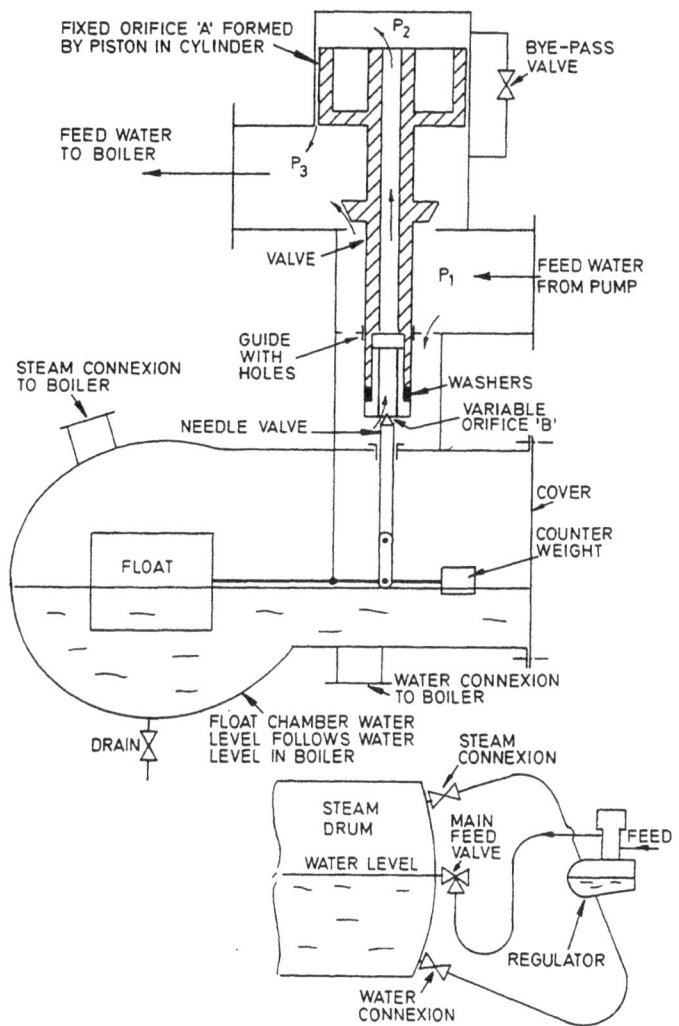

Fig. 1.33 BOILER FEED REGULATOR (WEIRS)

Operation of the regulator is as follows: Some of the feed water from the pump passes through the variable orifice 'B' (formed by the needle valve and the hole in the valve spindle) up to the piston chamber *via* the hole through the valve spindle and then to the boiler feed line through the fixed orifice 'A' (formed by the piston in the piston chamber). When the valve is in equilibrium the pressure

$$P_2 \simeq (P_1 + P_2)/2 \qquad P_1 > P_2 > P_3.$$

If the water level in the boiler rises, the water level in the float chamber will rise and the needle valve will be lowered, thus increasing the size of the orifice 'B' and increasing pressure P_2. The valve is then forced down until equilibrium is once again restored and a reduced quantity of feed water will now pass to the boiler.

If the water level in the boiler falls, the water level in the float chamber will fall and the needle valve will be raised, thus reducing the size of the orifice 'B' and reducing pressure P_2. The valve is then forced up until equilibrium is restored and an increased quantity of water will now pass to the boiler.

Washers are fitted to the valve spindle that enable the working water level to be altered.

Washers in—increased working water level

Washers out—reduced working water level in the ratio of 1 : 7 *i.e.* washer thickness: water level alteration.

For hand regulation of the feed water a by-pass valve is fitted. If this by-pass is opened the pressure difference P_2—P_3 is nearly zero and the feed valve will open to its full extent. Regulation of the feed water will then have to be carried out on the boiler feed check by hand.

An arrangement of this regulator for more sensitive control is shown in Fig. 1.34. This two-element regulator controls the feed water level by signals received from two sources, (1) Water level, (2) Steam flow. The construction of this regulator is similar to the single element type Fig. 1.33 except for the additional linkage to the needle valve from a specially spring-loaded stainless steel bellows that has its bellows space connected to the superheater outlet.

Main control of the feed supply is by means of the bellows unit. As the steam demand increases, the pressure in the bellows chamber will decrease, the needle valve will be raised and the feed supply to the boiler will increase. The reverse being the case for reduction in steam demand. With this arrangement a steadier water level is maintained in the boiler for varying steam demand.

If however changes in water level for some reason occur, the water

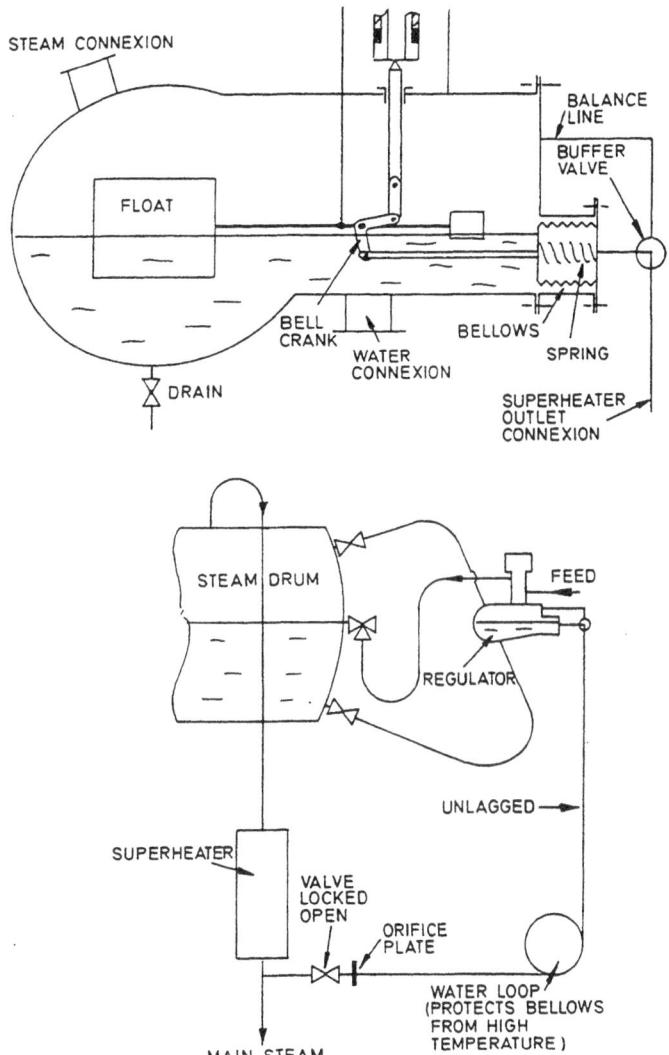

Fig. 1.34 TWO ELEMENT REGULATOR (WEIRS)

level control (i.e. float system) comes into operation over-riding the steam control.

To minimise water leakage into the superheated steam line in the event of the bellows failing, an orifice plate is provided. Also to safeguard the bellows from excessive pressure difference (it could not withstand boiler pressure on one side only) when taking the regulator out of service a buffer valve is fitted which would communicate the inside of the bellows with the float chamber pressure via the balance line.

Copes feed water regular shown in Fig. 1.35 is also of the 'two element type.' One is a steam flow element which senses pressure drop across the superheater, this will increase as steam flow increases. The other is a thermostat which responds to change in water level.

The thermostat consists of a polished stainless steel tube, fitted in a rigid steel frame, connected at its upper end by a lagged pipe to the steam space of the boiler drum, and by a unlagged pipe at its lower end to the water space.

Water in the lower part of the thermostat tube cools, by radiation, to a temperature below that of the steam in the upper part, hence in water level in the tube causes it to expand or contract.

With a rising water level the thermostat tube contracts and since the frame is clamped (as shown) to the tube, the lever will be rotated anti-clockwise. Its motion is relayed through the bell crank to the pilot valve. This alters the air supply pressure to the controller. The higher the water level the higher, in proportion, is the air supply pressure to the controller.

The two signals, one from the level sensing element the other from the steam flow sensor, are fed into a controller which relays the resultant signal to the valve positioner of the feed control valve. This valve is moved when the variables discussed alter.

An example of what could happen is as follows: If the burners are suddenly extinguished the resulting collapse of steam bubbles would lower the boiler water level, opening up the feed valve, the water level would then be restored and when the boiler is flashed up the water level would become dangerously high. However with the controller fitted this will not happen, since if the burners are extinguished steam demand must have fallen, this means that the controller will close in the feed valve reducing water flow to the boiler. For further discussion see control chapter.

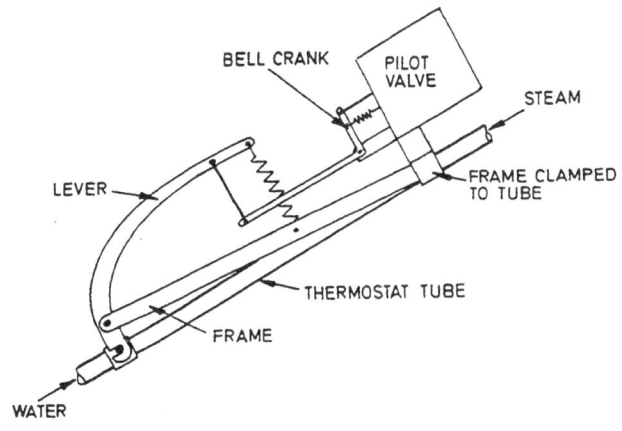

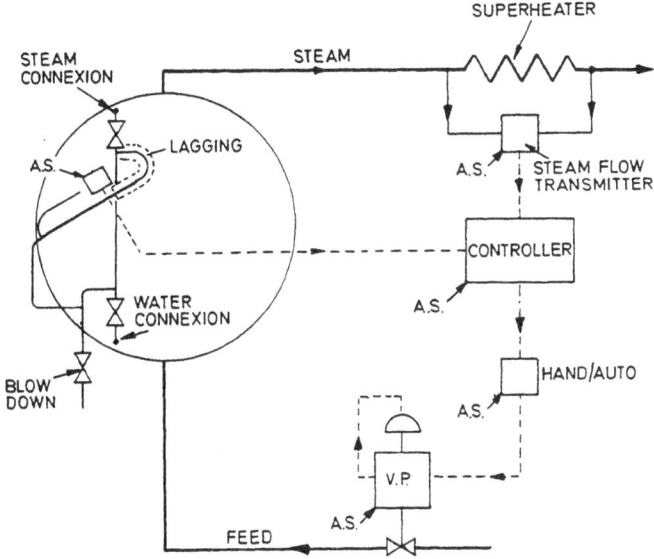

Fig. 1.35 COPES FEED WATER REGULATOR

Safety valves

The improved high lift safety valve and the full bore safety valve relay operated, are dealt with in Vol. 8 of the series since they are general questions. However, for water tube boilers operating at high pressures and temperatures the thermodisc type of safety valve with reduced blow down is finding increasing favour.

Blow down is important in the operation of the boilers, the term blow down refers to the steam wastage that occurs after the excess pressure is relieved and is caused by the valve not seating rapidly. If blow down can be reduced in any way then the plant is being operated more efficiently.

Fig. 1.36 shows some of the details of the Consolidated thermodisc type of safety valve suitable for high pressure operation (64 bar and above if required).

With the valve in the open position as shown in the main diagram steam is deflected as it escapes to give improved reaction effect on valve and piston assembly, this gives increased lift. Steam will also be passing through the holes in the valve assembly to chamber A and then to the atmosphere *via* the open vent.

When the valve starts to close down, the ventilation to atmosphere of the steam from chamber A will be stopped. This causes a pressure build up in the chamber which assists closing of the valve. The additional closing force due to this arrangement would be approximately equal to the pressure in chamber A multiplied by $(D^2 - d^2)\pi/4$.

In addition to this arrangement the reaction effect due to the blow down ring will be reduced as the valve is closing. The combination of these two adjustable items helps close the valve rapidly once the excess pressure has been relieved thus reducing blow down.

A thermodisc valve seat is fitted to the valve assembly and it is a relatively thin valve section. Due to its flexibility, and reduced possibility of thermal distortion due to temperature differences through the valve, the valve should give good steam tightness.

The reason for the thermodisc is basically steam tightness. If we consider a ordinary valve with a tiny steam leakage. The local area at the point of leakage is cooled due to steam throttling and the metal contracts. This contraction increases the gap between valve and seat, increasing the leakage rate. This effect is more pronounced at higher pressures *e.g.* saturated steam throttled from 100 bar to atmospheric pressure drops in temperature by 200° C approximately.

The thin thermodisc element (detail Fig. 1.35):

1.   Reduces thermal gradient across the seating surfaces.

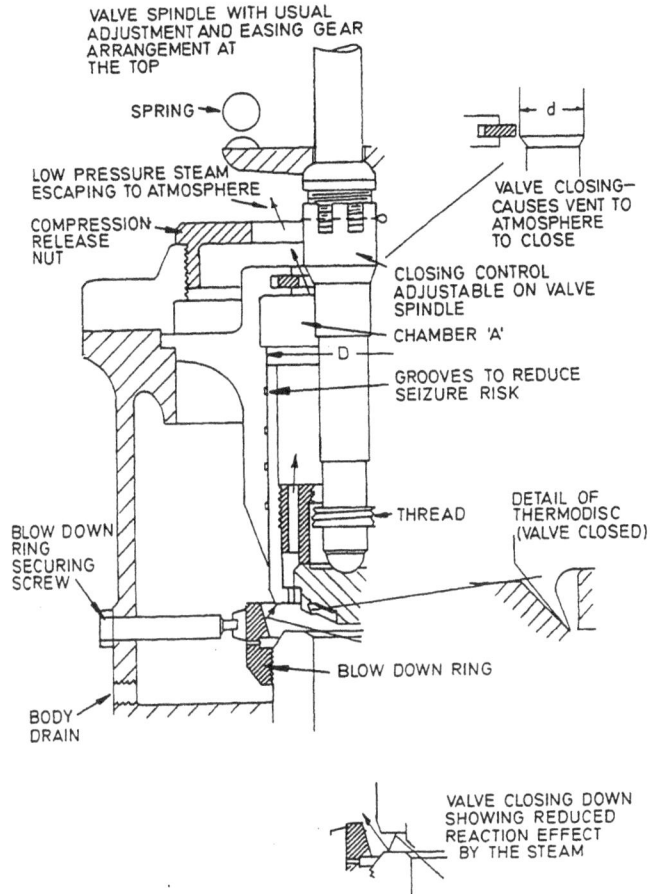

VALVE SPINDLE WITH USUAL ADJUSTMENT AND EASING GEAR ARRANGEMENT AT THE TOP

SPRING

LOW PRESSURE STEAM ESCAPING TO ATMOSPHERE

COMPRESSION RELEASE NUT

VALVE CLOSING— CAUSES VENT TO ATMOSPHERE TO CLOSE

CLOSING CONTROL ADJUSTABLE ON VALVE SPINDLE

CHAMBER 'A'

GROOVES TO REDUCE SEIZURE RISK

THREAD

BLOW DOWN RING SECURING SCREW

DETAIL OF THERMODISC (VALVE CLOSED)

BLOW DOWN RING

BODY DRAIN

VALVE CLOSING DOWN SHOWING REDUCED REACTION EFFECT BY THE STEAM

Fig. 1.36 SAFETY VALVE FOR HIGH PRESSURE BOILERS

2.   Deflects easily and accommodates any distortion of the seat.

3.   Enables good steam tightness due to steam pressure acting on the inner surface of the element forcing it onto the seat.

WATER TUBE BOILER EXAMINATION

The classification societies require all boilers where working pressure is in excess of 4.5 bar and where heating surface is greater than 4.5 m² to be examined every two years until the boiler is eight years old and every year thereafter.

An examination of the boiler must cover the external and internal parts including superheaters, economisers, air preheaters, casings, etc. Boiler mountings must be examined and safety valves tested and adjusted if required. The oil fuel burning system must be examined and tested. If necessary, plate or tube thicknesses must be checked and the boiler pressure parts may be tested hydraulically.

Before commencing to examine a boiler it is advisable to examine the boiler plans and make notes dealing with the route to be taken and the various items that should be examined along the route. Points to observe, with explanatory notes, along a typical route could be as follows:

1. Age of boiler

If boiler is old then previous surveys will have been carried out and possible repairs done. Familiarise oneself with the history, it could give indications of persistent faults.

2. Pressure and steam temperature

If the steam temperature is high at outlet from the superheater then special steel and welding techniques may have been used, if the boiler is new it is well to be aware of how these new materials, etc, are behaving in practice.

3. Inside steam drum

All the internal fittings, including those removed for access purposes, should be examined for corrosion and erosion. The attemperator (or desuperheater) tubes, if fitted, should be carefully examined since if they fail, water could enter the superheater causing overheating and failure of superheater elements. If considered necessary the attemperator may be pressure tested hydraulically.

Drum and tubes should be examined for corrosion and if any doubt exists regarding the internal condition of the tubes some of them should be removed (fire side tubes), cut up, and examined for scale and corrosion. Fire side tubes are chosen since these would be affected first.

Boiler mounting openings should be checked and boiler door tried into position.

4. Outside steam drum

All boiler mountings must be examined for evidence of leakages and corrosion. Check to ensure the mountings are securely fastened.

5. Back and side of boiler

Examine the inside of the superheater headers (if they are at the back) for scale and corrosion. Repeat the examination on the upper and lower headers for water walls. Carefully examine casing for leakages since furnace gas leaking into the boiler room could make conditions untenable. Check safety valves and drains.

6. Furnace

Examine the refractory for defects, the water drum is protected from excessive temperatures by refractory and if this refractory is damaged, cracking of the drum due to flame impingement could result, since the drum plating is relatively thick and therefore not effectively cooled.

Examine tubes for distortion, the roof tubes are nearly horizontal and in a region of high temperature so they may be among the first to be affected. Check superheater tubes and supports for distortion and wastage, excessive slagging (due to poor fuel or combustion) can also lead to tube wastage.

Where tubes enter drums check for leakages.

7. Inside water drum

The inside of the water drum must be examined in the same way as the inside of the steam drum. Check blow down valve.

8. Superheater—gas side

The boiler is usually designed so that the gas side of the superheater can be readily examined. This is important since it operates at the highest temperature. Slagging of the superheater can lead to high furnace pressures because of the obstruction leading to inefficient boiler operation and where gases can pass through, overheating of the superheater elements can occur.

Superheater supports have often burnt away in the past due to ineffective or non-existent cooling. If the elements are not supported then additional stresses are set up in the tubes, drainage is affected and the gas flow is altered possibly causing a reduction in boiler efficiency.

9. Economiser and air preheater

The internal and external condition of the tubes must be ascertained, externally the tubes may have been affected by acid attack, internally the economiser tubes may be scaled up or corroded.

EXTERNAL BOILER CLEANING

1. Soot blowing

Regular use of soot blowers is essential if the boiler is to be maintained in a clean and hence efficient condition. The time period between operations of the blowers will be governed by the installation requirements, *e.g.* if a poor fuel is being used or if the vessel is operating on short voyages with long standby and manoeuvring periods.

In a water tube boiler the soot blower operating sequence would generally be (after increasing fan pressure to help remove deposits by increased gas flow).

1.   Operate air preheater or economiser blowers since this is where the greatest soot deposits are likely to be encountered.
2.   Operate furnace soot blowers.
3.   Operate economiser then air preheater soot blowers in order to chase the deposits from the furnace up the uptake and out of the boiler plant.

2. Water washing

Bonded deposits on superheaters, economisers and air preheaters may be encountered, especially when using the poorer grades of fuels, fortunately these bonded deposits, which generally cannot be removed by soot blowing, contain water soluble substances. By using a solution containing a wetting agent (to reduce the surface tension) and an alkali in conjunction with common salt to soak the deposits with before using the hot, high pressure water jets, the removal of the deposits can be greatly simplified.

Washing should be done from the top of the boiler working down and should be continuous. Drains are provided in the furnace floor and these can be located by inspection of the boiler plans. After washing all the deposits out of the boiler it should be thoroughly dried by using a small fire.

WATER TREATMENT

Boiler water treatment has been dealt with in detail in volume 8 of the series, hence only recently adopted and proposed developments will be considered here together with some of the new topics raised in examination questions.

To produce water suitable for the modern high pressure boiler and to maintain the contamination of the water to a minimum once it is in the system requires a relatively sophisticated production and treatment plant. A modern production and treatment plant could be made up of the following:

1. Distillation plant

Various types are available but most modern installations use low pressure evaporation (including flash) in order to produce distillate with low solids content as cheaply as possible.

A well designed evaporator will be capable of producing distillate with total dissolved solids content of less than 1 p.p.m. but with sufficient carbon dioxide to give a pH value of around 5.4. This acid distillate can corrode ferrous and non-ferrous metals.

Methods of minimising and possibly eliminating this corrosion are:

1. Add volatile amines to raise pH value.
2. Use, ideally, non corrodible metals.
3. Protect make up feed storage tanks with the addition of small amounts of polyphosphate to the distillate and pre-coat tanks with protective paint.
4. Prevent undercooling of evaporator distillate. Carbon dioxide solubility increases with decrease in temperature.
5. Use demineralisation plant.

2. De-mineralisation plant

The distillate from the evaporation plant will contain salts, these can be completely removed in a de-mineralisation plant. Fig. 1.37 shows diagrammatically such a plant, it operates on the ion exchange principle. In each of the exchangers is a resin which replaces the salts with ions as indicated.

In order to appreciate better what is happening in an ion exchanger the way in which elements are bonded together will be briefly examined. There are two main types of chemical bonding called

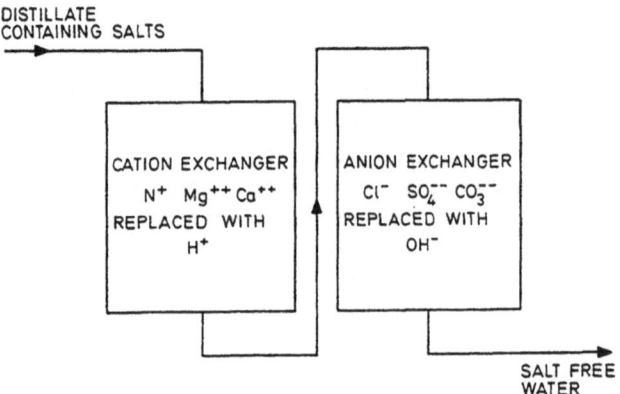

Fig. 1.37 DE–MINERALISATION PLANT (ION EXCHANGER)

covalent bonding and electrovalent bonding.

Covalent bonding: Electrons are shared between elements in order that the elements may be bonded together to form compounds, an example is carbon dioxide CO_2, here the electrons of the oxygen and carbon elements are shared.

Electrovalent bonding: Electrons are given and received between elements in order to produce a bonded compound, an example is sodium chloride $NaCl$, here the electron donor is the Na and the electron acceptor is the Cl, hence $NaCl = Na^+ + Cl^-$. The donor and acceptor are ions since they have gained or lost electrons. They have electrical charge and this is indicated by the signs (which also indicate how many electrons have been gained or lost by the elements, one sign one electron, two signs two electrons, etc.) the Na^+ is positively charged and is called a cation, the Cl^- is negatively charged and is called an anion.

It is also possible to obtain compounds which act as anions, *e.g.* SO_4^{++} and CO_3^{--}.

Hence in the ion exchanger the cations are replaced with hydrogen ions from the exchange resin and the anions are replaced with hydroxyl ions. The resins have to be regenerated by using an acid and an alkali to replace the hydroxyl and hydrogen ions and to wash out the salts.

Fig. 1.38 shows a mixed bed de-mineralisation plant of the regenerative type.

When the resins require regenerating, indicated by a predetermined maximum conductivity reading e.g. 0.5 μmhos/cm^2, the following sequence is used.

1. Backwashing with fresh water. Due to anion resin density being lower than the cation, backwashing causes separation of the resins.

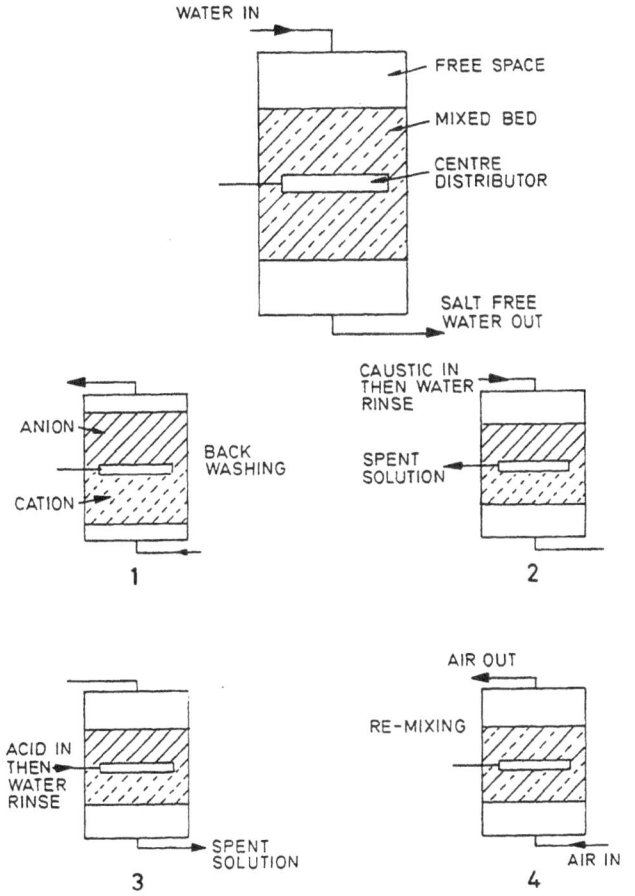

Fig. 1.38 MIXED BED DE–MINERALISATION

The free space in the chamber allows for expansion of the bed during this process.

2.   Regeneration of the anion resin with a caustic solution to replace the OH^- ions. Follow with a water rinse to remove spent solution.

3.   Regeneration of the cation resin with a acid solution to replace the H^+ ions. Follow with a water rinse.

4.   Re-mix the bed by blowing air through.

A typical unit could handle up to 10 tonnes/h and could purify 1800 tonnes of water before requiring regeneration—this depends upon purity of water supplied.

The trend today is to use an ion exchange plant with a single exchanger (or de-mineralisation bed as they are often called) wherein the resin is compound and is replaced with a new charge after a certain period of time has elapsed—this depends upon the condition and quantity of feed passing through.

3. De-aeration plant

The water from the de-mineralisation plant is make up and reserve feed for the feed system, the water already in the feed system can become contaminated with gases.

Possible sources of contamination are:

(a)   undercooling of the condensate in the condenser.
(b)   bled steam supply pressure to the de-aerator falling off,
(c)   dirty air ejector filters,
(d)   ingress of small amounts of salt water.

De-aeration would be twofold mechanical and chemical.

After mechanical de-aeration—achieved by boiling the feed and removing the non condensible gases, the oxygen content may be as low as 0.005 p.p.m. and the carbon dioxide 0.001 p.p.m. But even these small amounts of gases can cause corrosion hence it is necessary to treat the feed chemically.

It is worth noting that one of the main sources of trouble in modern high pressure boiler plant is corrosion of the feed and condensate system which results in copper-iron scale being picked up and deposited in the boiler. These deposits set up corrosion cells, due to dissimilar metals in boiler water, which causes loss of tube material.

A fact that has emerged is the copper-iron scale composition is remarkably consistent, being approximately 70% iron, 10% calcium,

10% copper, remainder made up of zinc, magnesium and sodium. It is possible that the scale deposited on water wall tubes could cause the tubes to become insulated. This would lead to overheating of the metal below the scale and water being flash evaporated, leaving boiler water solids on the metal surface in a highly concentrated form. This is turn could lead to caustic attack.

Hydrazine solution (60% Hydrazine 40% water approximately) is finding increasing popularity for oxygen scavenging, it reacts under boiler conditions with the oxygen to form water,

reaction: Hydrazine + Oxygen $\longrightarrow$ Water + Nitrogen
$$N_2H_4 + O_2 \longrightarrow 2H_2O + N_2$$

thus having the advantage of not increasing the boiler water density.

Initially it was thought that excessive dosage of hydrazine could lead to steam and condensate line corrosion due to ammonia being produced as the excess hydrazine decomposed:

(Hydrazine $\longrightarrow$ Ammonia + Nitrogen)

However, a *controlled* excess is beneficial to the steam and condensate system as it counteracts the effect of carbon dioxide corrosion. This figure would be approximately 0.1 to 0.3 p.p.m. of hydrazine.

Hydrazine should be injected after the de-aerator pump discharge, this is logical for several reasons:

1.   The de-aerator is being allowed to do the job it is designed for.
2.   De-aerator performance can be checked by taking a water sample before hydrazine injection point.

Excess hydrazine can be converted according to the equation

Hydrazine $\longrightarrow$ Ammonia + Nitrogen
$$3N_2H_4 \longrightarrow 4NH_3 + N_2$$

The nitrogen and ammonia carry over with the steam from the boiler and when the steam condenses the ammonia reacts with the water to form ammonium hydroxide.

i.e. Ammonia + Water $\longrightarrow$ Ammonium Hydroxide
$$NH_3 + H_2O \longrightarrow NH_4OH$$

As the ammonium hydroxide is alkaline it will increase the pH of the condensate and if oxygen is present it can lead to corrosion of copper alloys.

There may be a delay in the build up of a reserve of hydrazine in the boiler water since it reacts with any metal oxides (apart from Fe_3O_4) that may be present.

Sodium sulphite may still be used as an oxygen scavenger, if that is the case then the following points regarding it are important: (a) pH value is important to reaction rate with the oxygen, at about 7 pH it is a maximum hence the sodium sulphite should be injected into the system before any alkaline ingredients, (b) In high pressure boilers the sulphite can break down to give hydrogen sulphide (H_2S) and possibly sulphur dioxide (SO_2) which can attack steel, brass and copper, (c) It increases dissolved solids content.

4. Condensate line treatment

Where the steam is wet, and also in the condensate system, corrosion can occur due to the presence of carbon dioxide carried over with the steam. To ensure alkalinity in this section of the system a volatile alkaliser may be injected into the steam line. These alkalisers are generally ammonia or cyclohexylamine, they combine with the steam as it condenses to form carbonates and bicarbonates which decompose in the boiler to give back the CO_2 and the alkaliser, some of which then returns to the steam system.

If the pH value of the condensate is maintained at about 9 this should ensure no corrosion in the low temperature steam and condensate sections of the plant.

Such a pH value would not be maintained with hydrazine alone, the previously mentioned monocyclohexalamine with morpholine can be added (at any point in the system, they distil off with the steam and re-circulate continuously) to maintain correct feed pH.

When the boiler is new, silica (SiO_2) can concentrate in the water (even after good pre-commissioning). Silica can volatilise and carry over with the steam and deposit upon turbine blades, affecting efficiency. By keeping pH carefully controlled the amount of silica the water contains can be controlled, the higher the pH the more silica the water can contain and the greater the carry over.

Semi automatic and fully automatic water treatment systems are available.

With the semi automatic system the operator analyses the make up feed and the boiler water, he then doses the system and operates blow down by remote control or by preset timing devices, the additives are injected in solution form.

A fully automatic system will require a computer to give continuous analyses of make up feed, condensate and possibly water in the boiler. It will then adjust the continuous treatment of the water and blow down if required.

With modern high pressure automated boiler plant intermittent analysis of the water is not satisfactory. The main reasons are:

1. The high water purity required.
2. The monotony of a very frequent essential manual and mental exercise, i.e. testing and analysing.
3. The extremely high evaporative rate of the boiler, which in a relatively short space of time could lead to instigation of damage if water condition deteriorates.

Fig. 1.39 shows a continuous monitoring arrangement for some of the boiler water parameters. The monitoring devices have their signals relayed to a control room where they can be under continuous supervision.

Conductivity: this measures salt concentration in the water.

Hydrazine: this measures hydrazine reserve which is essential for maintaining the black magnetite (Fe_3O_4) layer, removing oxygen and keeping the water in a reducing condition.

Signals from the phosphate (PO_4) and pH monitors can be fed into a X–Y plotter for comparison with e.g. the standard graph used for the co-ordinated phosphate—pH treatment. Fig. 1.39 shows the form of graph used with this treatment. Simply, if the intersection of the co-ordinates lies below the curve there will be no free caustic in the boiler water.

It is possible to convert the signals received in order to operate chemical dosage equipment.

Note. The treatment of the boiler water with phosphates and coagulants, etc, has been dealt with in Volume 8 so it was not included in the foregoing discussion of a modern treatment plant although it would be an integral part of such a plant.

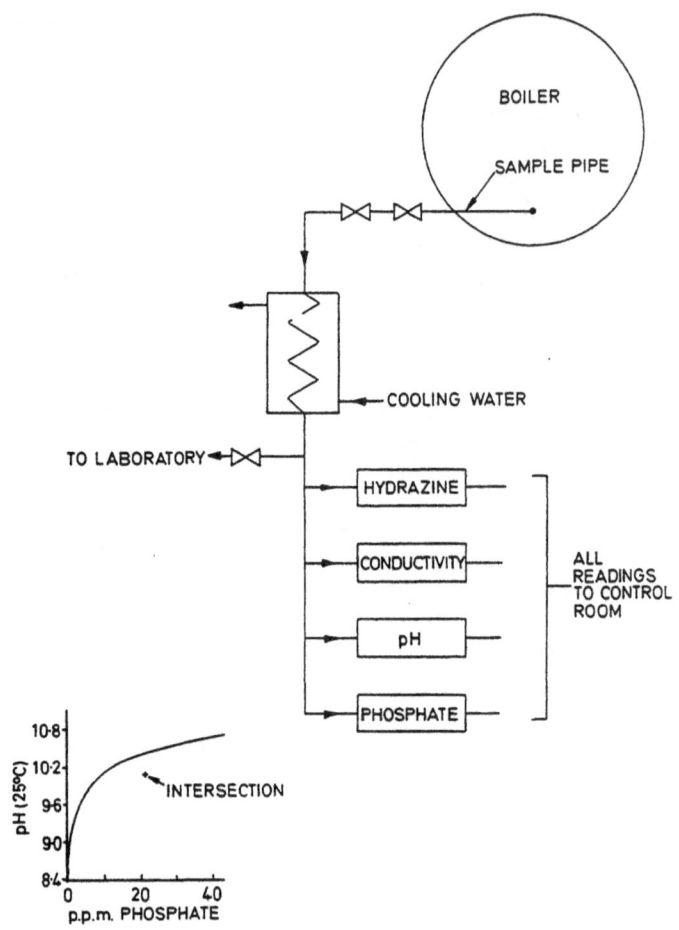

Fig. 1.39 CONTINUOUS BOILER WATER ANALYSIS

Conductivity measurement

Specific conductivity mho/cm^3 is the conductance of a column of mercury 1 cm^2 cross sectional area and 1 cm long. This is a large unit and micromho/cm^3 is used and when corrected to 20° C is called a dionic unit.

A conductivity meter is usually an ordinary ohmeter, calibrated in conductivity units, with a small hand driven generator across two electrodes which are adjustable in distance apart to allow for temperature correction.

The electrodes are immersed in a tube containing the water sample and after temperature adjustment the reading is obtained by rotating the generator for a few revolutions and noting the scale reading of conductivity.

Conductivity of pure distilled water 0.5 dionic units.

Conductivity of fresh water is about 500 dionic units.

Total dissolved solids measurement

This can be evaluated on a rough first basis from the conductivity meter provided that chemical treatment is first applied. Conductivity is mainly due to OH^- (hydroxyl) and H^+ (hydrogen) ions so that these must be first standardised for all solutions before the actual dissolved solids reading can be correlated from the conductivity reading.

Standardisation usually consists of adding acid to the boiler water (assumed alkaline) to neutralise hydroxyl ions. Conductivity is then on a common basis due to dissolved mineral solids in the water, which constitute the dissolved solids.

A calibration scale change graph or alternative instrument scale is necessary to give a dissolved solids reading. The sample may require dilution with distilled water to reduce conductivity to a readable value.

Electrical salinometer

The electrical supply to the instrument must be direct current. The potentiometer is provided to give a fixed standard calibration voltage so that no errors to voltage differences exist.

Pure water is non conducting so that current flow is an indication of impurities, i.e. the greater the current the greater the impurity in the water.

Referring to Fig. 1.40.

When the impurity content exceeds a fixed value the current is sufficient to operate the Relay 2 so giving visual or audible warning by closing the circuit. Continued operation at increasing current would cause Relay 1 to short circuit the meter and so protect it.

Water temperature increases conductivity so that a temperature compensation is required. A temperature compensating thermometer

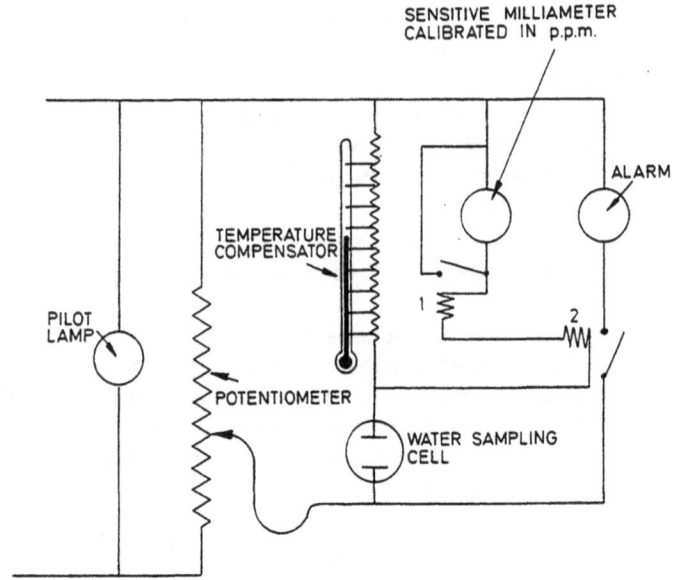

Fig. 1.40 ELECTRICAL SALINOMETER

is as a shunt across the meter. Temperature increase causes a rise of mercury level and a cutting out of resistance which allows more current through the shunt and less through the meter. The correct calibration current through the meter is fixed, current variations due to temperature are shunted.

TEST EXAMPLES

Class 2

1.   Make a detailed sketch of a main boiler having a roof fired furnace showing the gas path through the boiler. State one advantage and one disadvantage of roof firing. Explain how gas flow is achieved.

2.   Make a detailed sketch of a superheater showing its position in a main boiler and the flow paths of steam and combustion gases.

Describe three defects commonly found in superheaters and explain their cause.

3. Sketch and describe an ion exchange column. State with reasons where it is situated in the plant. Explain its function.

4. Describe with sketches how a burst tube in the membrane wall of a main boiler furnace is dealt with. Give two advantages of membrane wall construction. Define the importance of correct membrane · width and the effect of too broad or narrow a membrane.

5. Describe with sketches one way whereby steam from a main boiler is maintained at approximately constant temperature irrespective of generation rate.

State what vital precaution needs to be observed when raising steam in a boiler possessing this facility.

State why auxiliary steam is not usually taken from the steam drum.

6. Draw in detail a Cope's boiler feed regulator, labelling all components.

State with reasons what advantages it has over the simple float type.

Explain with sketches how:

(a) the thermostat element responds to water level changes,
(b) the steam flow element responds to steam flow changes.

Class 1

1. Sketch the furnace of a radiant heat boiler showing the disposition of tubes and headers and direction of water flow. Comment on tube diameter, pitch, and material used. State how the generating tubes are attached to the headers. Give two reasons why unheated down-comers are sometimes fitted. State why radiant heat boilers are claimed to be superior to most other designs.

2. With reference to main boilers:

(a) sketch the various ways whereby superheater elements are attached to the headers.
(b) suggest with reasons which method in (a) is superior to the others.

3. Explain how steam delivered by a main boiler may be maintained at an approximately constant temperature independent of the generation rate by:

(a) the position of the superheater,

(b) the use of mechanical control of gas flow,

(c) means external to the boiler.

Describe one arrangement to supply low superheat steam by means external to the boiler.

Explain why the supply is not taken directly from the steam drum.

4. Explain with sketches how the following characteristics are incorporated in high capacity spring loaded safety valves:

(a) opening quickly and fully without wire drawing,

(b) closing quickly with equal precision,

(c) complete tightness until moment of lift.

5. Draw in detail a feed regulator whose action is only partially governed by boiler water level.

Explain how it functions.

State with reasons what advantages it possesses over other types of regulator.

6. With reference to boiler water purity and treatment write notes on each of the following:

(a) magnetite,

(b) hydrogen embrittlement,

(c) neutralising amines,

(d) filming amines.

CHAPTER 2

TURBINE THEORY

Steam turbines can be broadly classified into two groups, *impulse* and *reaction.*

In the simple impulse turbine steam is supplied to a set of fixed nozzles wherein the steam falls in pressure and gains in velocity. The high velocity steam issuing from the nozzles is directed to impinge upon a row of blades which are fixed to the periphery of a rotor, this causes the rotor to be rotated.

In the reaction turbine, alternate rows of fixed and moving blades, attached to the casing and rotor respectively, form passageways which serve as fixed and free nozzles. The steam passes through the fixed blading and gains in velocity, it then impinges upon the moving blades to give an impulsive effect. The steam then passes through the moving blades expanding as it does so to give a reaction effect—hence this type of turbine is often referred to as impulse-reaction.

The foregoing is just a brief outline of the impulse and reaction turbines which will be examined in greater detail in this and the following chapter.

TURBINE NOZZLES

In the design of turbine nozzle many factors have to be taken into consideration, some of the important ones are:

1. The velocity of steam at exit from the nozzles that is required.
2. The shape of the nozzle.
3. The quantity of steam that has to pass through the nozzle per unit time.
4. The temperature of operation.

1. The velocity of the steam at exit from the nozzle is generated by

converting heat energy into kinetic energy.

If we assume the velocity of the steam approaching the nozzle is small in comparison to the velocity of the steam leaving the nozzle then it may be neglected. The velocity of the steam leaving the nozzle may then be calculated from the following formula:

$$V = \sqrt{2gH}$$

where H is the heat drop of the steam as it passes through the nozzle in J/kg.

where g is the acceleration due to gravity, *i.e.* 9.81 m/s^2.

and V is the leaving velocity in m/s.

Hence the desired velocity of the steam leaving the nozzle is governed by the heat drop, which in turn is governed by the pressure drop. The pressure drop is dependent upon the shape of the nozzle.

2. The shape of the nozzle as has been stated above governs the velocity of the steam issuing from the nozzle.

If the velocity V is required to be *subsonic* then a converging type of nozzle will be used. If however V has to be *supersonic* the nozzle generally used is of the converging-diverging type, better known as the De-Laval nozzle.

As nozzle length increases the losses due to friction also increase, for a nozzle about 50 mm length this may be about 5% of the heat drop in the nozzle and for 150 mm length about 15%. Convergent nozzles are usually small in length, convergent-divergent have to be relatively longer in order to obtain a large exit velocity at low pressure.

In the convergent-divergent nozzle the divergence must not be too great as this means the outer steam streams are not at a favourable angle for the blades, these outer steam streams will produce shock at entry to the blades with subsequent vibration effects, hence the divergence angle is kept to about 12°. To limit the adverse effect and waste of the outer steam streams, a parallel extension may be provided (see Fig. 2.1). It may be thought that the divergence angle could therefore be increased in order to reduce nozzle length and friction loss. However, if the divergence angle is made too great (20° or more) the shock at the change in section (divergent to parallel) results in turbulence, that will effect the steam streams and reduce nozzle efficiency. It must be remembered that adding length to the nozzle, by the introduction of a parallel section, increases the friction loss and this must be balanced with the gain in having parallel steam streams.

3. The mass of steam that passes through the nozzle per unit time is dependent upon (a) the velocity of the steam at the throat, (b) the area of the throat and (c) the specific volume at the throat (*i.e.* the m^3/kg of steam).

If m = kg of steam passing through the nozzle every second.

A = area of the throat in m^2.

V = velocity of the steam at the throat in m/s.

v_2 = specific volume of the steam at the throat in m^3/kg.

v_1 = specific volume of the steam at the nozzle
  entrance in m^3/kg.

p_1 = pressure of steam at nozzle entrance in N/m^2.

p_2 = pressure of steam throat in N/m^2.

Hence $m = VA/v_2$ kg/s   1.

also $p_1v_1^n = p_2v_2^n$ is the expansion law from entrance to throat.

$$\therefore \left(\frac{p_1}{p_2}\right)^{\frac{1}{n}} v_1 = v_2 \text{ substitute into equation} \qquad 1.$$

$$\therefore m = \frac{VA}{v_1} \times \left(\frac{p_2}{p_1}\right)^{\frac{1}{n}} \qquad 2.$$

It can be shown (but it is beyond the scope of this book) that equation 2 is a maximum when $p_2/p_1 \simeq 0.57$, the actual ratio depends upon the index n which can vary from 1.3 for superheated steam to about 1.113 for very wet steam.

The pressure $p_2 \simeq 0.57p_1$ is called the critical pressure. It is the pressure at which A is a minimum and the discharge per unit area is a maximum.

At the critical pressure the steam will have sonic velocity, that is the maximum velocity that can be obtained in a convergent nozzle. If A is reduced beyond the minimum necessary to obtain the critical pressure, no further increase in velocity will be obtained but the discharge rate from the throat will be reduced.

The velocity of the steam can be increased beyond the sonic velocity obtained at the throat of the convergent part of a nozzle by adding a divergent portion. Steam issuing from the throat into the divergent section behaves like a controlled expanding projectile where final velocity is dependent upon the pressure drop from throat to outside of nozzle, (p_2 to p_3).

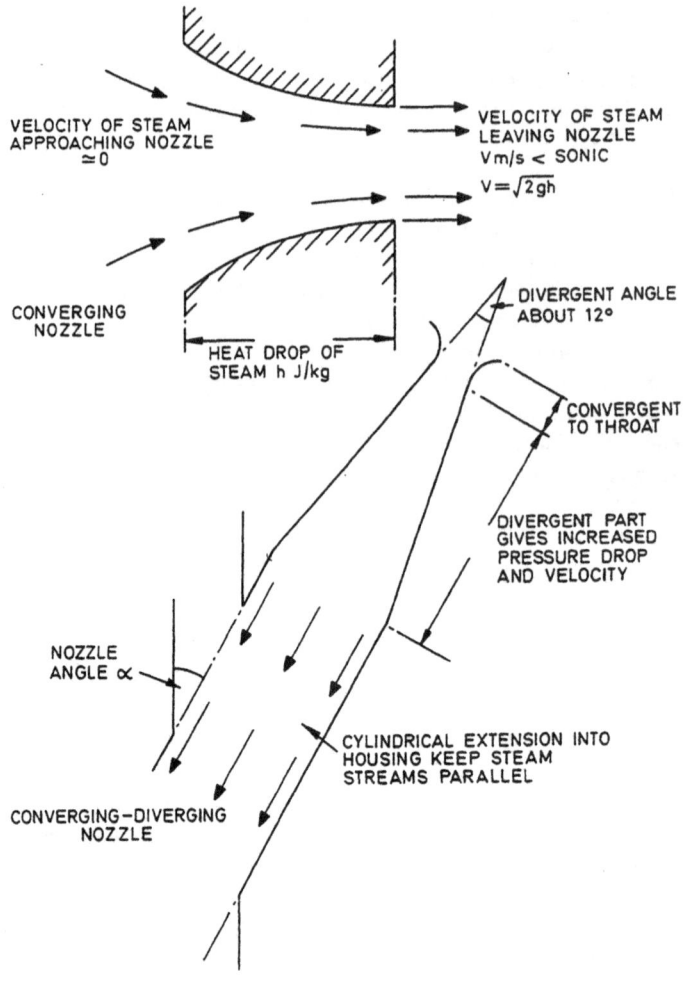

VELOCITY OF STEAM
APPROACHING NOZZLE
$\simeq 0$

VELOCITY OF STEAM
LEAVING NOZZLE
V m/s < SONIC

$V = \sqrt{2gh}$

CONVERGING
NOZZLE

HEAT DROP OF
STEAM h J/kg

DIVERGENT ANGLE
ABOUT 12°

CONVERGENT
TO THROAT

DIVERGENT PART
GIVES INCREASED
PRESSURE DROP
AND VELOCITY

NOZZLE
ANGLE α

CYLINDRICAL EXTENSION INTO
HOUSING KEEP STEAM
STREAMS PARALLEL

CONVERGING–DIVERGING
NOZZLE

Fig. 2.1

Fig. 2.2 shows variation of quantity discharge with variation of critical pressure ratio and inlet pressure. m/A is the kg of steam passing through the throat per second per unit area of throat.

p_2/p_1 is the pressure ratio throat to inlet.

Curves 1 and 2 show how m/A increases as p_2/p_1 is reduced, when

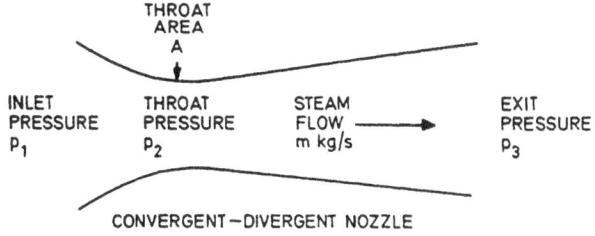

CONVERGENT – DIVERGENT NOZZLE

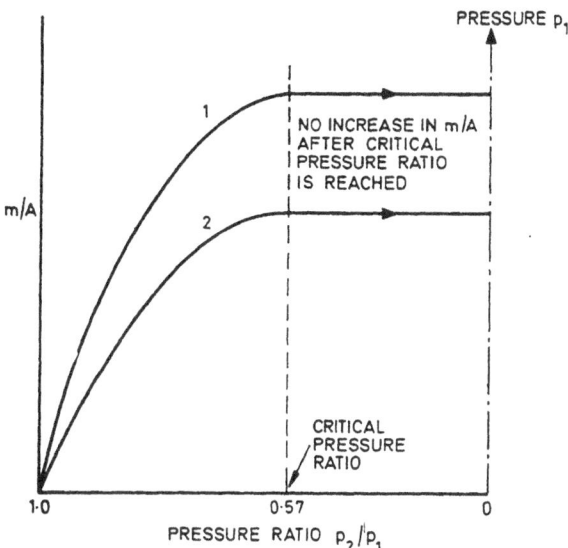

Fig. 2.2

$p_2 \simeq 0.57p_1$, m/A is at a maximum. Any further reduction in p_2/p_1 does not increase the discharge rate.

Curve 1 is for a lower inlet pressure p_1 than curve 2.

4. The temperature of operation (and the pressure) governs the choice of material to be used for the nozzles.

Materials used must be:

(a) Of high strength to withstand the high pressures.

(b) Hard enough to withstand the erosive effect of the steam.

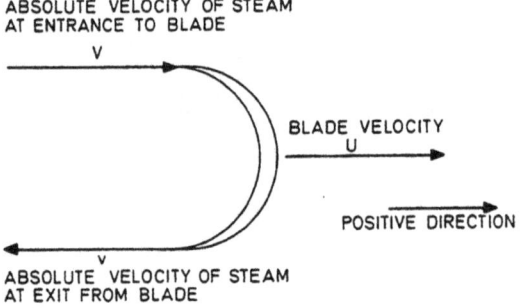

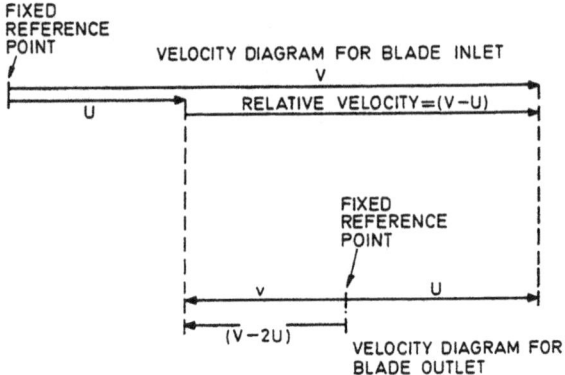

Fig. 2.3

(c)   Creep resistant in order to maintain its shape.

(d)   Resistant to corrosion.

Stainless steel—stainless iron or monel metal would be a probable choice.

SIMPLE IMPULSE TURBINE STAGE EFFICIENCY

In the ideal simple impulse turbine consisting of nozzles and a single row wheel, the steam would issue from the nozzles in the direction of motion of the blades (this cannot be so in practice). Fig. 2.3 shows the arrangement with steam leaving the nozzles at velocity V and blades moving with velocity U.

It will be clear from the diagrams of velocity that the change in absolute velocity of the steam in the direction of motion of the blades (this direction being positive) is given by

$$V - [-v]$$
$$i.e. \quad V - [-(V - 2U)]$$
$$i.e. \quad 2(V - U)$$

If m is the kg of steam entering the blade per second, then from Newtons second law the force on the blade is

$$F = m2(V - U) \text{ N}$$

and the power developed $P = m2(V - U) \times U$ Nm/s
The power available in the steam $= \frac{1}{2}mV^2$ Nm/s

Stage efficiency

$$\eta = \frac{\text{power developed}}{\text{power available}}$$

$$\eta = \frac{m2(V - U)U}{\frac{1}{2}mV^2}$$

$$\eta = \frac{4U(V - U)}{V^2}$$

$$\eta = 4\left[\frac{U}{V} - \left(\frac{U}{V}\right)^2\right] \qquad 1.$$

Fig. 2.4 is a graphical illustration of equation 1 and it will be clearly seen from the graph and table of values that maximum stage efficiency occurs when the blade velocity is exactly one half of the steam velocity at inlet to the blades. With this ratio, *the absolute velocity of the steam at exit from the blades must be zero.*

It would be of interest at this stage to compare the operating torque at peak efficiency with the torque exerted under standing conditions.

If R = mean blade ring radius.

Torque $T = F \times R$
$$T = m2(V - U)R$$

VELOCITY RATIO U/V	0	1/8	1/4	3/8	1/2	5/8	3/4	7/8	1
EFFICIENCY η	0	7/16	3/4	15/16	1	15/16	3/4	7/16	0

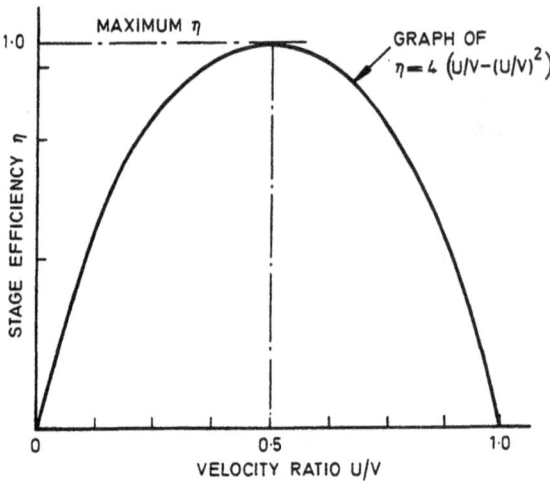

Fig. 2.4

$$\therefore T = m2VR(1 - U/V) \qquad 2.$$

When $U = 0$, $T = m2VR$

When $U = V/2$, $T = mVR$

Hence the standing torque is theoretically twice the value of the torque obtained under maximum efficiency running.

Fig. 2.5 shows graphically equation 2.

ACTUAL VELOCITY DIAGRAMS FOR AN IMPULSE STAGE

Fig. 2.6 shows the actual impulse turbine arrangement of nozzle to blade. Steam leaving the nozzle cannot enter the blading tangential to the blade ring, however the nozzle angle a is usually kept as small as possible (10° to 20°) to reduce side thrust on the blade ring.

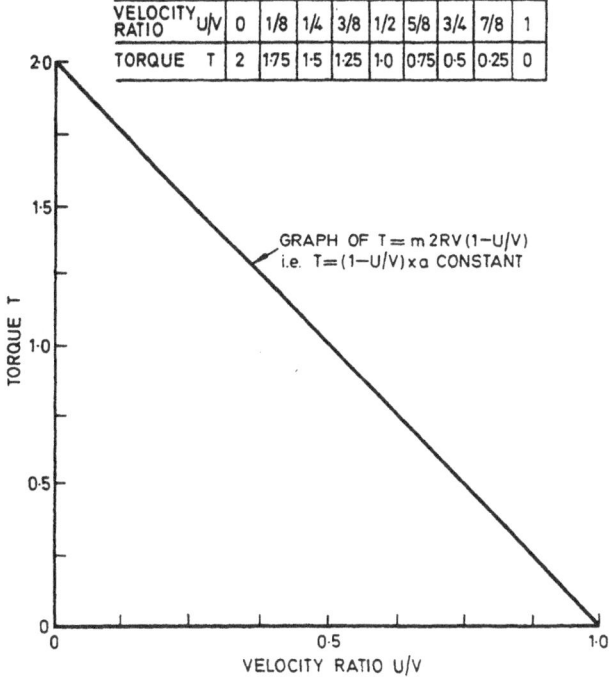

VELOCITY RATIO U/V	0	1/8	1/4	3/8	1/2	5/8	3/4	7/8	1
TORQUE T	2	1·75	1·5	1·25	1·0	0·75	0·5	0·25	0

GRAPH OF $T = m\,2RV(1-U/V)$
i.e. $T = (1-U/V) \times a$ CONSTANT

TORQUE T

VELOCITY RATIO U/V

Fig. 2.5

The blade inlet angle θ_1 must be chosen so as to ensure under normal operating conditions of running that the steam enters the blading without shock (i.e. enters smoothly). From the velocity diagram at blade inlet it can be seen that the relative velocity of steam to blade at inlet, V , makes an angle with the blade ring of θ_1 thus ensuring smooth entry.

The change in absolute velocity of the steam in the direction of motion of the blades is given in this case by:

$V \cos \alpha - (- v \cos \beta)$ positive direction again to the right.

i.e. $V \cos \alpha + v \cos \beta$ this expression is given the symbol V_w and is called the *velocity of whirl.*

Power developed would be $mV_w U$ Nm/s

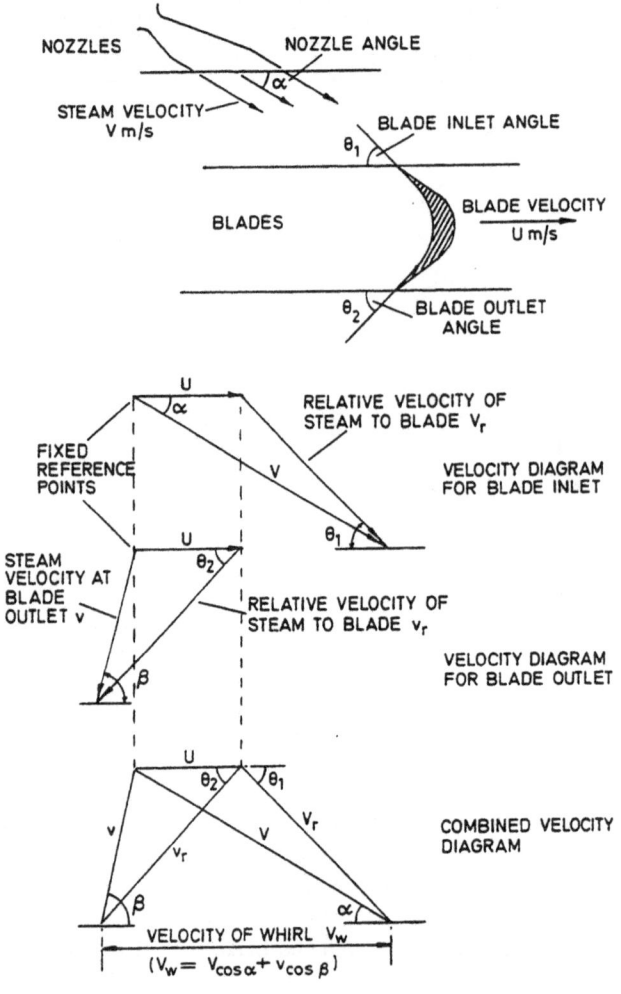

Fig. 2.6

If we assume no friction losses through the blades and that the path through the blading is parallel there will be no change in relative velocity of the steam to the blade.

i.e. $V_r = v_r$

Also if blading is assumed symmetrical $\theta_1 = \theta_2$

$$\therefore \quad V_r \cos \theta_1 = v_r \cos \theta_2$$
$$\therefore \quad V \cos a - U = v_r \cos \theta_2$$
$$\therefore \quad v \cos \beta = V \cos a - 2U$$
$$\text{and} \quad V_w = V \cos a + v \cos \beta$$
$$\therefore \quad V_w = 2(V \cos a - U)$$

Stage efficiency $\eta = \dfrac{\text{power developed}}{\text{power available}}$

$$\eta = \frac{m2 \ (V \cos a - U)U}{\tfrac{1}{2}mV^2}$$

$$\eta = \frac{4(V \cos a - U)U}{V^2}$$

$$\eta = 4 \left[\frac{U}{V} \cos a - \left(\frac{U}{V} \right)^2 \right] \qquad 3.$$

If this equation is compared with equation 1,

$$\eta = 4 \left[\frac{U}{V} - \left(\frac{U}{V} \right)^2 \right]$$

it differs only by the cos a term. For a nozzle angle $a = 20°$ the maximum efficiency that can be obtained is approximately 90% with U/V being approximately one half. As a is increased the efficiency of the stage will be decreased, this is the important fact derived from equation 3.

In practice the stage efficiency will be lower than that obtained from equation 3 because of (a) Friction loss, (b) Steam leakage, (c) Windage and pumping losses.

At optimum velocity ratio it would possibly be of the order of 80%, see Fig. 2.7.

COMPOUNDING FOR VELOCITY

A velocity compounded section of a turbine consists of a row of nozzles, through which a large pressure drop takes place, followed by alternate rows of moving and fixed blades. Theoretically, the path

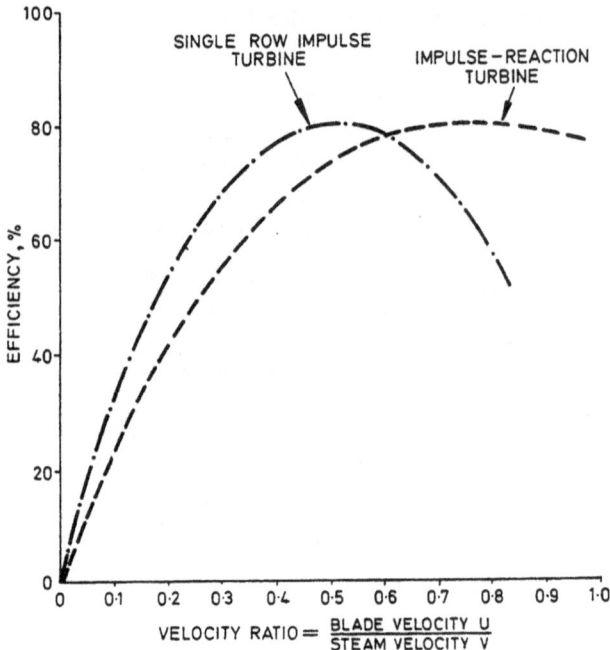

Fig. 2.7

through the moving and fixed blades is kept parallel so that the pressure remains uniform as the steam passes through.

The moving blades are attached to a wheel (so forming what is often called a Curtis wheel) which in turn is fixed to the rotor shaft or is forged integral with the rest of the rotor. The fixed blades are attached to the casing and they need only extend over an arc sufficient to cover the arc formed by the nozzles. In some turbines the nozzles may be arranged all around the casing, the fixed blading would then have to extend all around the casing. This arrangement gives a uniform force distribution on the wheel, which would help to reduce vibration.

Advantages to be gained through using a two or three row wheel velocity compounded are:

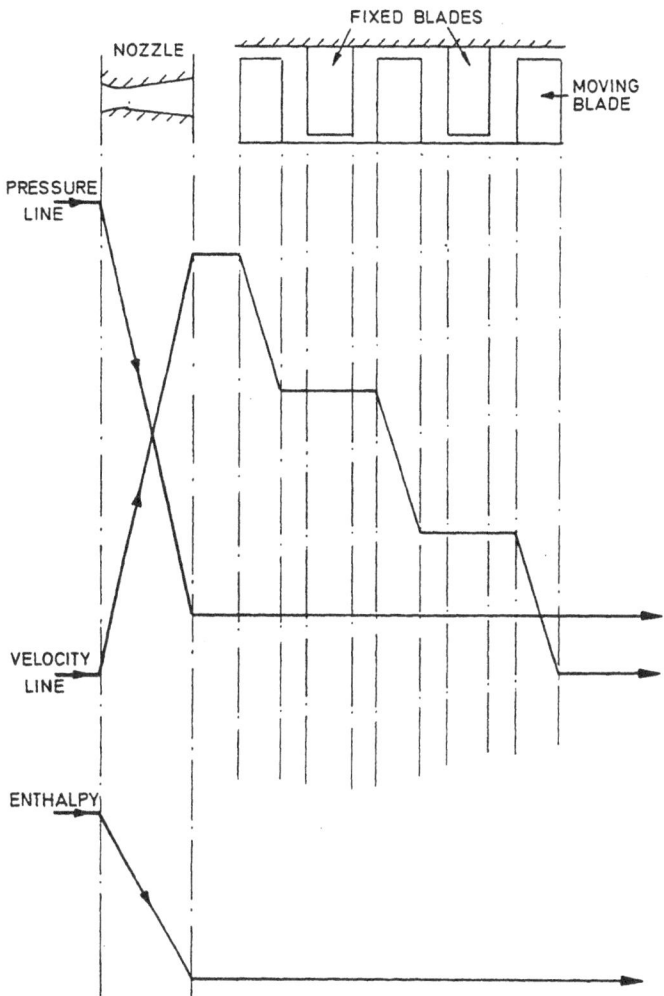

Fig. 2.8 COMPOUNDING FOR VELOCITY

1. Large pressure drop is obtained through the nozzle which has the following effects due to small casing pressure:

(a) Glands less difficult to keep steam tight.

(b)   Gland length can be reduced.

(c)   Reduced pressure stresses.

2.   Reduction in turbine length.

3.   Reduction in cost since less material would be required, also the casing material does not have to withstand high temperature creep conditions so less expensive metal can be used.

The disadvantages to be weighed against these advantages are:

1.   Lower efficiency.

2.   Increased steam consumption.

Maximum theoretical efficiency for
velocity compounded wheel

Referring to Fig. 2.9 and to the velocity diagrams given in Fig. 2.3 the absolute velocity of the steam at exit from a row of moving blades in the simple tangential entry impulse turbine is given by subtracting twice the blade velocity from the absolute velocity of the steam at inlet. This means therefore that the leaving velocities would be:

For a single row $V - 2U$
For a double row $(V - 2U) - 2U = V - 4U$
For a triple row $(V - 4U) - 2U = V - 6U$.

The fixed blading just diverts the path of the steam, theoretically without velocity loss.

It was shown (equation 1, page 87) that the maximum efficiency for a single row of moving blades occured when the velocity ratio $U/V = \frac{1}{2}$, *i.e.* when the blade velocity equalled one half of the steam velocity at inlet and that the *absolute velocity of steam at exit was equal to zero.*

If we now considered a two row velocity compounded wheel (*i.e.* two moving rows). Maximum efficiency would be obtained when the leaving velocity was zero.

$$i.e. \; V - 4U = 0$$

In this case the blade speed must be equal to $V/4$ and the theoretical efficiency would be 75% (substitute U $V/4$ into equation 1, page 87).

Now consider a three row velocity compounded wheel. Maximum

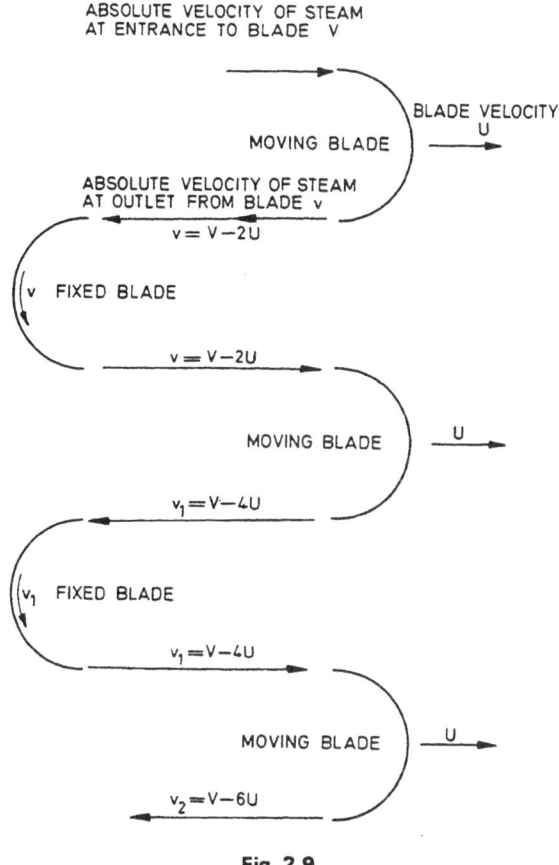

ABSOLUTE VELOCITY OF STEAM
AT ENTRANCE TO BLADE V

BLADE VELOCITY
U

MOVING BLADE

ABSOLUTE VELOCITY OF STEAM
AT OUTLET FROM BLADE v

$v = V - 2U$

v FIXED BLADE

$v = V - 2U$

MOVING BLADE   U

$v_1 = V - 4U$

v_1 FIXED BLADE

$v_1 = V - 4U$

MOVING BLADE   U

$v_2 = V - 6U$

Fig. 2.9

efficiency again would be obtained when the leaving velocity is zero.

i.e. $V - 6U = 0$

Blade speed would now be equal to $V/6$ and the theoretical efficiency would be 55% approximately.

When practical considerations are taken into account these stage efficiencies would probably be more like:

65 to 70% for a two row wheel,

and 50% for a three row wheel.

In an actual velocity compounded section such as that shown in

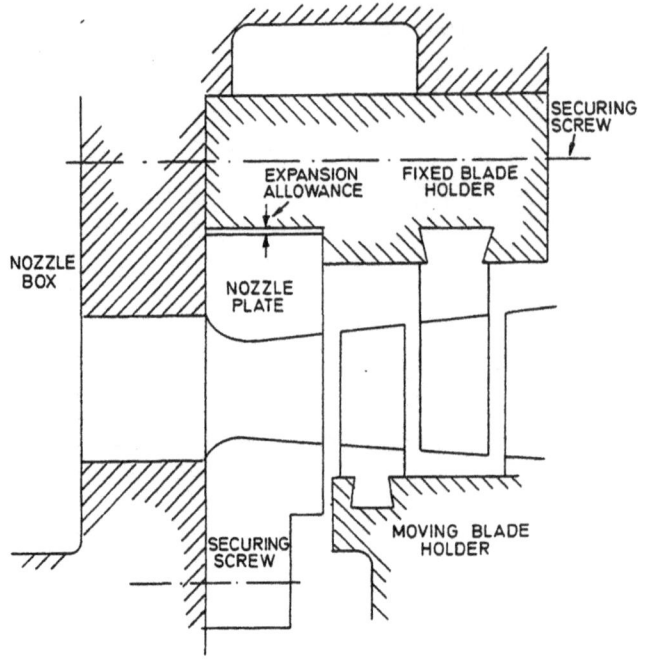

Fig. 2.10 VELOCITY COMPOUNDED SECTION FOR IMPULSE TURBINE

Fig. 2.10 the nozzle box and fixed blade holder will extend over the same arc. Blading will gradually increase in height as we move away from the nozzle since the velocity of the steam is falling off, mass flow is constant and the specific volume of the steam is to all intents and purposes constant (some slight variation does take place in the specific volume of the steam due to reheating caused by friction).

As different materials are used in the construction of the nozzle plate and casing, and temperature differences are encountered, expansion allowances have to be made.

THEORETICAL CONSTRUCTION OF AN IMPULSE TURBINE BLADE

Profile blading is mainly used in impulse turbines, having super-ceded the now rather old fashioned plate type blade. Profile blading is robust and efficient, usually the blading pathway will be non parallel so that a certain reaction effect is produced in addition to the impulse obtained from the steam issuing from nozzles.

Fig. 2.11 is a step by step construction for a very simple equiangular parallel path profile blade. Referring to the diagram we commence the construction by 1. Drawing three equally spaced parallel lines, A, B and C. 2. Draw triangle a b c with blade inlet angle, for constructional convenience, made equal to $30°$ *i.e.* $\theta_1 = 30°$. 3. Draw arcs of radius r_1 and r_2 using c as centre where $r_1 = cb$ and $r_2 =$ distance from c to line ab. 4. Construct another triangle a b c with apex c at centre of arc radius r_2. Now draw common tangents to the arcs in order to complete the profile. In the diagram, the various stages of the construction have been done by means of separate diagrams, the student is advised to construct the blade profile by superimposing the separate diagrams one upon the other.

A modern standard impulse blade is also shown in Fig. 2.11. The inlet edge has in practice to be given some thickness in order that the mechanical strength of the blade is not impaired, however, this edge can cause steam flow difficulty at inlet and to reduce the difficulty and improve efficiency the edge is rounded. An extension of the blade at exit helps to give better guidance to the steam as it leaves the blade, again improving upon efficiency.

With a modern impulse blade an increase in efficiency of 5 to 15% can be obtained over the older symmetrical type the actual value would depend mainly upon the inlet angle to the blade.

COMPOUNDING FOR PRESSURE

A pressure compounded impulse turbine is made up of a number of stages. Each stage consists of a row of nozzles fixed to a diaphragm followed by a row of blades attached to a wheel. In other words it is built of a number of simple impulse turbines.

The nozzle carrying diaphragms are fitted into the casing in halves and between diaphragms a uniform pressure zone exists in which the blade carrying wheel, either fitted to the shaft or forged integral with the shaft, rotates.

To prevent steam leakage between the uniform pressure zones, each zone being at a lower pressure than the previous one, shaft seals or diaphragm packing is provided.

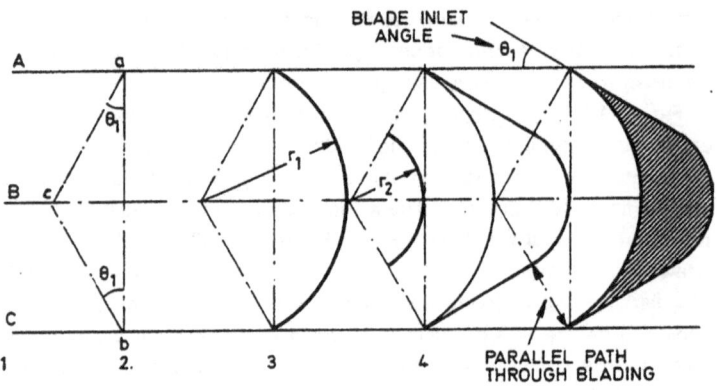

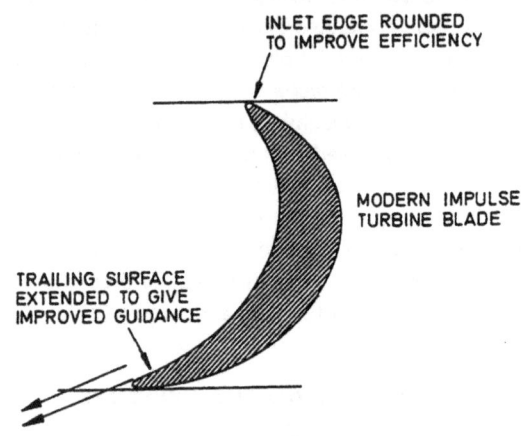

Fig. 2.11

Advantages of pressure compounding are:

1.   Steam velocities are lowered due to the smaller heat drop that occurs through the diaphragm nozzles. With smaller blading velocity, slower speeds of rotation are possible and centrifugal effects are minimised. (It must be remembered that centrifugal force is proportional to the *square* of the velocity.)

2. Efficiency is increased. If we wish to obtain the same reduced blade velocity in a velocity compounded wheel with one row of nozzles that would be obtained in a pressure compounded turbine. Then a drop in stage efficiency must be accepted due to the three or four row velocity compounding required. Whereas stage efficiency is high for each single impulse turbine unit used in a pressure compounded system.

The principal disadvantages of pressure compounding would be:

1. Increased first cost.
2. Increased length of turbine.

Various combinations of velocity compounding and pressure compounding have been used in order to try and obtain the benefits to be gained from each individually. A common arrangement is to have a two row velocity compounded section followed by a pressure compound section, this gives a neat, compact, efficient and relatively moderately priced turbine.

REACTION TURBINE

In the introduction to this chapter a brief description was given of the reaction turbine principle, to take this further, reference can be made to Fig. 2.13.

This Fig. shows the path of the steam formed by the aerofoil shaped blading attached to casing and rotor respectively. Casing blading passes all round the turbine and the moving blades are attached to a cylindrical rotor, generally of uniform diameter throughout its length being stepped down at its ends to form bearing journals. Both fixed and moving blades form nozzles, one row of fixed blades and one row of moving blades forming a stage—hence there are equal numbers of rows of fixed and moving blading.

As the steam passes through a row of fixed casing blades it falls in pressure. The heat drop through the blading results in an increase in absolute velocity of the steam. Use is made of the change in velocity in the direction of motion of the moving blades to give an impulsive effect.

The moving blades are identical in shape to the fixed blades, the path through them again being made convergent so that the steam will fall in pressure. There will be, due to the fact that the blades are

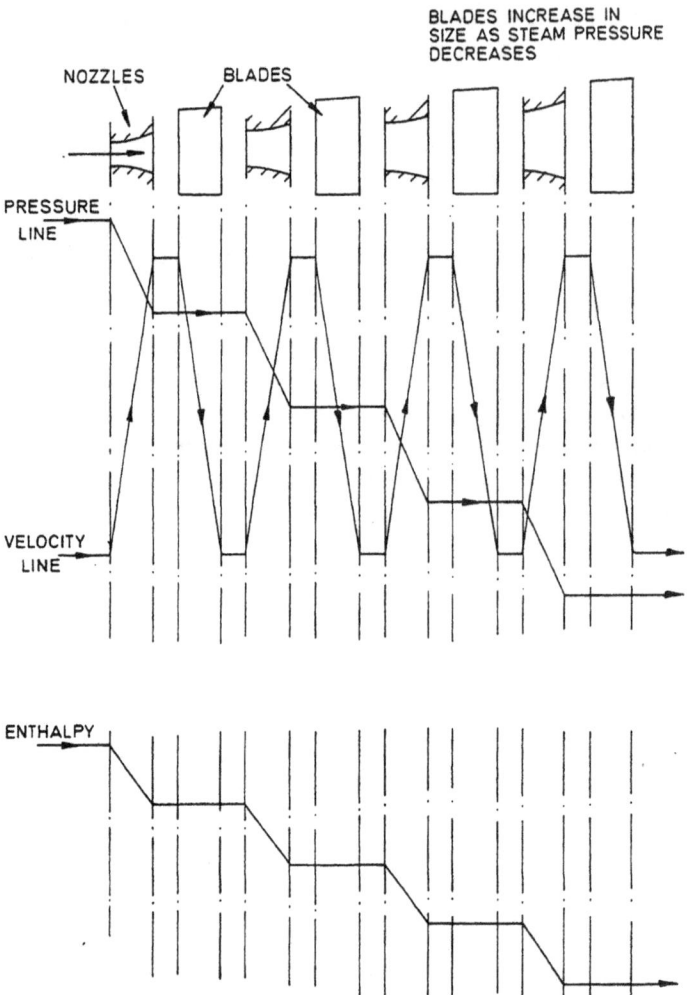

Fig. 2.12 COMPOUNDING FOR PRESSURE

moving, a fall in the absolute velocity of the steam but a gain in relative velocity will be achieved. A reaction occurs due to the change of velocity *within the blading*—this is a free nozzle effect. This turbine therefore is often referred to as the impulse-reaction instead of more simply a reaction turbine.

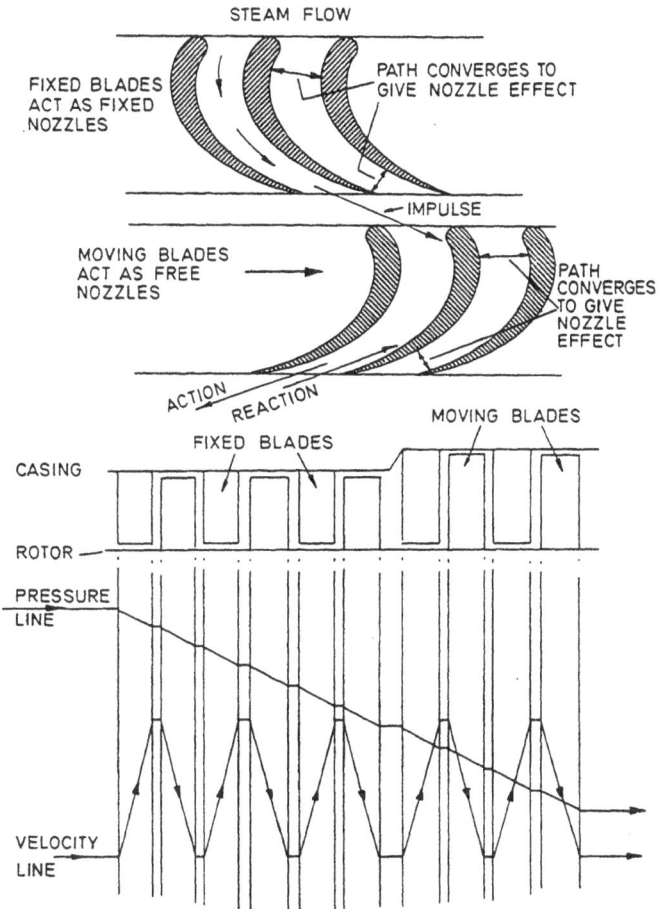

Fig. 2.13 REACTION TURBINE

Theoretically, as the steam pressure falls through the turbine and the specific volume increases, each stage should increase in height to accommodate the increase in volume. However, this would prove expensive on a large turbine with a great number of stages, so a compromise is sometimes reached by stepping the casing and having two, three, four or more stages of equal heights (see Fig. 2.13).

It will be appreciated that blade running clearances must be kept small in order to limit the loss of steam that could pass the blading to perform useful work, this loss would be greatest at the higher pressures due tc the low specific volume. In order to limit loss at high pressures the impulse-reaction turbine could be preceded by a velocity compounded impulse wheel. This would have the added effect of reducing turbine length and number of reaction stages required.

Reaction turbine velocity diagrams

With identical forms of blading for fixed and moving, ignoring friction losses and assuming perfect entry into moving blades, the inlet and outlet velocity triangles are similar such that:

$$V_r = v \text{ and } V = v_r$$
$$\theta_1 = \beta \text{ and } \alpha = \theta_2$$

The velocity of whirl $V_w = V \cos \alpha + v \cos \beta$

and the power developed in the stage is $mV_w U$ Nm/s.

In order to obtain maximum theoretical efficiency in an impulse-reaction turbine the blade velocity U should be equal to the steam velocity V, if entry was tangential to the blade ring. However, with entry as shown in Fig. 2.14 the blade velocity would be approximately 0.9 of the steam velocity. Fig. 2.7 shows how the efficiency of a stage in a reaction turbine varies with variation of velocity ratio U/V and it can be seen that maximum efficiency occurs in the region

$$0.8 \leqslant U/V \leqslant 0.9$$

Approximate construction for a reaction turbine blade profile Fig. 2.15

1.   Commence by drawing two parallel lines and let the perpendicular distance between the lines be unity (*i.e.* distance AB).
2.   Draw line AC at angle θ_2 (*i.e.* outlet angle) to line AB. $\theta_2 \simeq 15°$.
3.   With compasses set at distance AB centre C strike an arc across AB to obtain centre D.
4.   With compasses set at distance AB centre D draw arc CE.
5.   Extend BA to F such that $BF = 1\frac{1}{2} \times AB$.
6.   Draw line KG parallel to AE and with compasses set at distance BF centre F draw arc CG.

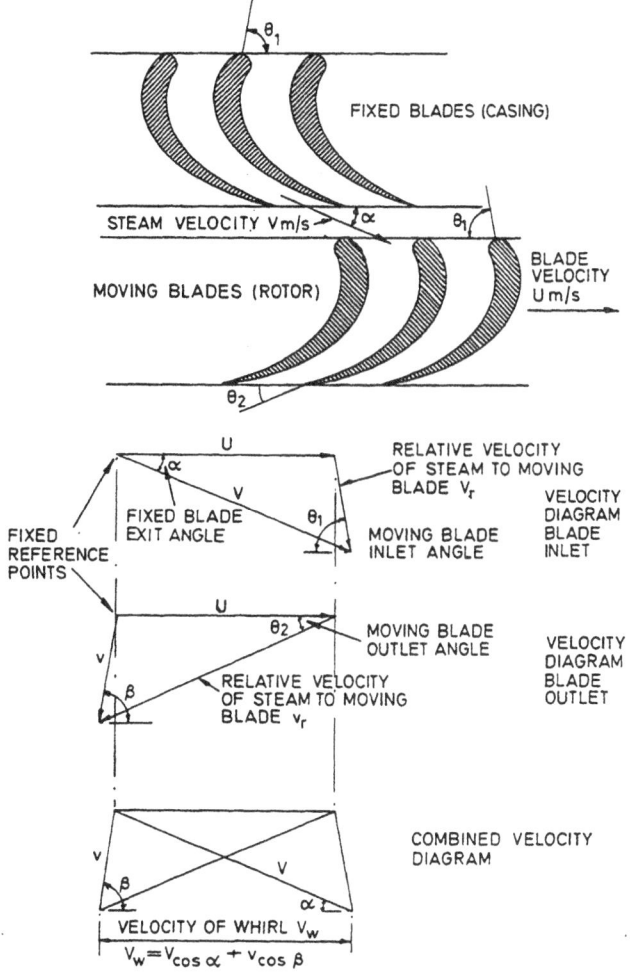

Fig. 2.14

7. With compasses set at distance $\frac{1}{2} \times$ AB centre G strike an arc, repeat using centre E so that intersection of arcs gives point H.
8. Using centre H draw arc GE.

GAS TURBINES

Gas turbines have not proved as extremely popular as it was first thought that they would. Initial development was rapid but when it was found that their efficiency was relatively low and little could be done to improve it, development slowed down considerably. The main difficulty it seems is to use the high gas temperature that can be obtained. If the temperature drop through the turbine can be increased then the efficiency will be increased, however, materials limit the inlet gas temperature to the turbine. Experiments have been carried out using ceramic materials for turbine blading so that the temperature at inlet could be increased from about 750° C to about 1200° C however the ceramics are invariably very brittle and difficult to form into accurately dimensioned shapes, hence more research will have to be carried out before such materials would be acceptable in practice.

Various blade cooling systems have been developed and these have proved more successful than using ceramic materials, however the cost of a cooling system can be high when one thinks in terms of hollow blading, etc.

A simple gas turbine consists of a compressor and turbine fitted on to a single shaft. The compressor delivers air at a pressure of about 4.8 bar and temperature 200° C to a combustion chamber in which the fuel is burnt. The combustion gases at a temperature of about 750° C leave the combustion chamber and enter the turbine, wherein, they fall in pressure and temperature thereby giving up energy. Exhaust from the turbine would be at about 450° C and atmospheric pressure.

Power output from the turbine is used to drive the compressor and the load, the compressor takes 60% or more of the power output.

Simple gas turbines of this nature would not be a practical marine proposition since they are not suitable for widely varying load conditions and the starting motor required would have to be enormous.

A separate turbine or *Load Turbine* as it sometimes is called overcomes the difficulties mentioned above. The compressor with turbine on the same shaft can be easily started with a relatively small starting motor—it would not have to turn gearing and propeller—and it can be kept running at a steady speed during manoeuvring conditions. Gas flow to the load turbine can be regulated in order to accommodate changes in power demand. The load turbine could drive the

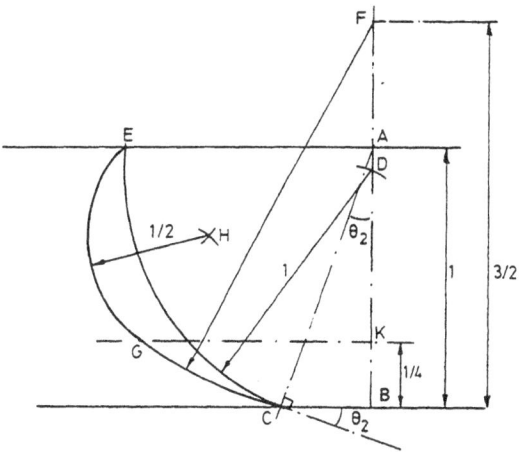

Fig. 2.15 APPROXIMATE CONSTRUCTION FOR A REACTION TURBINE BLADE

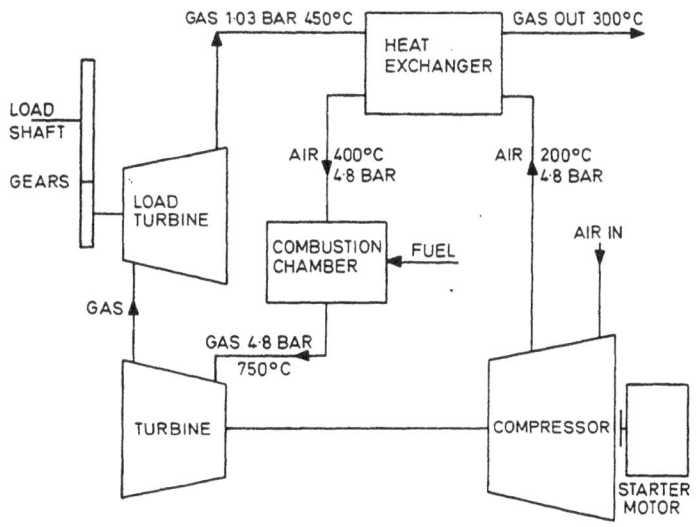

Fig. 2.16 GAS TURBINE WITH HEAT EXCHANGER

propeller through a clutch unit of the hydraulic type which would be reversible, thus no astern turbine would be required. If an astern turbine is fitted then a drop in efficiency would be automatic due to its friction, windage and pumping losses. An alternative to a clutch could be a controllable pitch propeller.

Such a turbine plant would not be very efficient, its fuel consumption would probably be in the region of 0.4 kg/kWh—compare this with a diesel engined installation whose fuel consumption could be 0.21 kg/kWh.

To improve on the efficiency, use can be made of the heat energy in the exhaust gases to preheat the air leaving the compressor. The arrangement is shown diagrammatically in Fig. 2.16, it could possibly reduce the fuel consumption to about 0.25 kg/kWh. An additional improvement could be to use multi-stage compression with intercooling between the stages.

Free Piston Gas Generators

The free piston gas turbine plant consists of one or more gas generators which supply gas to the turbine or turbines. A gas generator combines compressor and combustion chamber into one unit and consists of two piston units opposed to each other but operating freely. Each piston unit is made up of one large and one small diameter piston, the smaller piston reciprocates in a cylinder in which air inlet ports F, gas outlet ports and fuel injectors are provided. Chamber C is the cushion chamber, D and E air compression and receiver chambers respectively.

Cycle of operation

If we consider the pistons at the extreme outward position about to move inwards, chambers D and E filled with air at about atmospheric pressure and cushion air in chamber C at maximum pressure.

Moving inwards towards each other the large pistons will discharge air through valves B into chamber E raising the air in pressure. The small pistons meanwhile compress the air in the engine cylinder to a sufficiently high enough pressure and temperature to enable the fuel which is injected to be ignited and burnt.

When the pistons move from the extreme inward position outover, due to the hot high pressure gas in the engine cylinder, air will be

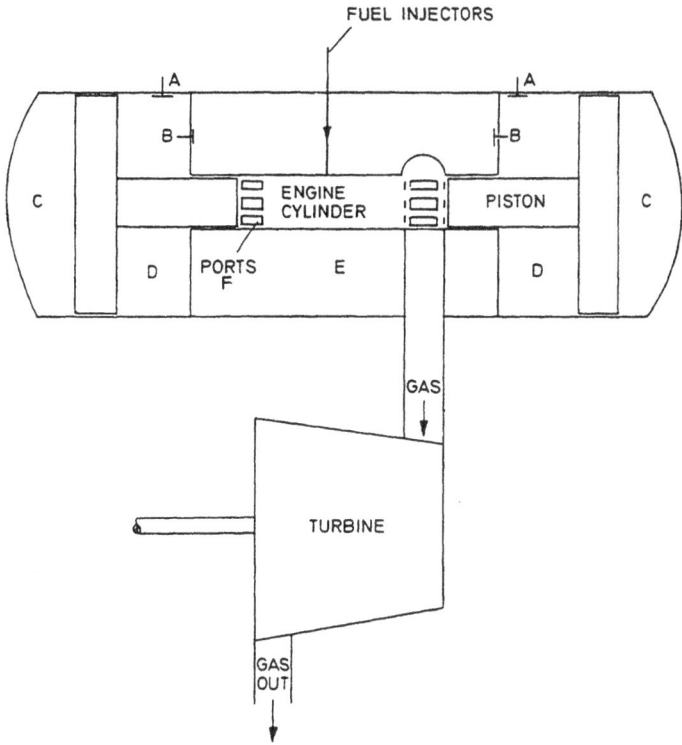

Fig. 2.17 FREE PISTON GAS TURBINE

drawn in through valves A onto chambers D, valves B closed and air in the cushion chambers C being compressed. A point will be reached on the outward stroke when the pistons start uncovering the gas ports and scavenge ports F. Then gas can pass to the turbine, and scavenging and recharging of the engine cylinder can take place with pressurised air passing from chamber E through ports F. When the pistons reach the extreme outward position the cycle of operations is back to the starting point.

Under steady operating conditions the equation:

Work done in engine cylinder = Work done in compressor cylinder holds good.

If the work done in the engine cylinder is increased by altering the fuel supply, then pressures will increase and the stroke will increase if the above equation is to be obeyed.

Increasing the pressure increases the frequency of operation (*i.e.* cycles/s [Hertz]) hence the gas quantity delivered per unit of time will increase.

Fuel timing must be advanced as the load is increased in order to maintain good combustion. It must be remembered that with a *free* piston unit stroke and compression ratio is a variable.

Starting of the gas generator is achieved by the use of starting air being supplied to the cushion chambers.

TEST EXAMPLES

Class 2

1. Explain with the aid of diagrams what is meant by (a) Velocity compounding. (b) Pressure compounding in impulse turbines.

2. Discuss with diagrams (where required) the passage of steam through a reaction turbine, explain clearly what happens to the steam as it passes through a stage in the turbine.

3. What are the differences between convergent and convergent-divergent nozzles? What are the effects of these differences?

4. Discuss the advantages and disadvantages of using a two or three row velocity compounded impulse wheel in a turbine.

5. Explain briefly a gas turbine system, its method of operation and its disadvantages.

Class 1

1. Discuss the various design requirements for the steam nozzles used in impulse turbines, explain how the steam increases in velocity as it passes through the nozzle.

2. Show graphically how the pressure and velocity of steam varies as it passes through, (a) a velocity compounded wheel (b) a pressure compounded turbine. What advantages are there in velocity compounding?

3. Compare impulse and reaction turbine principles stating where one would be more usefully employed than the other.

4. Make a diagrammatic sketch of a free piston gas generator. Explain its method of operation and enumerate the advantages to be gained by using a gas generator.

CHAPTER 3

GEARING

Various types of gears are used to produce a positive drive between two shafts which are next to one another. Those of major interest to the turbine engineer are the spur gears and helical gears.

Spur gears have teeth which are parallel to the shafts, and helical gears have each tooth cut on a helix. With parallel shafts engaged with helical gearing the term spur gearing may be used to describe the gears.

GEARING DEFINITIONS

Pitch cylinders:

Are the imaginary cylinders of the gears that could roll together without slip. The section of a pitch cylinder at right angles to the cylinder axis is called the *Pitch circle.*

Circular pitch:

This is the arc length on the pitch circle circumference between identical points on adjacent teeth.

Diametral pitch (d.p.):

Is the number of teeth per unit of diameter of the pitch circle.

$$i.e.\ d.p. = \frac{\text{Number of teeth}}{\pi \times \text{diameter of pitch circle}}$$

Addendum:

Is the radial height of tooth from pitch circle to tip circle.

Dedendum:

Is the radial depth of tooth from pitch circle to root circle.

The working contact surface of the addendum is called face and working surface of the dedendum is called flank. An all addendum gear is when the pitch circle radius is located at the outside diameter of the main gear wheel so that the wheel teeth have all dedendum flanks and the pinion teeth are all addendum faces. Such a gear gives relative sliding over the whole tooth surface, giving smooth running and reduced risk of pitting.

Tooth loading factor 'K'

The K factor, as it is generally called, is of importance to the gear designer and manufacturer but knowledge of its existence and approximate values gives the practising Marine Engineer a better appreciation of gearing. In the same way that the Engineer develops a sense of 'pH values' he can acquire a sense for the 'K' factor.

Evaluation of the factor is relatively complex, the revised formula given in Lloyd's register is $W = KdG/(G + 1)$ where W is the maximum tangential tooth load in force/unit length
where d is the pitch circle diameter
where G is the gear ratio, i.e. pinion speed/wheel speed.

Obviously if d and G are constant, the higher the values of K, the higher the value of W and hence the greater can be the power transmitted.

With unhardened steel pinions and wheels of the type used in the 1940–50 era K varied from about 40 to 80.

With through hardened steels K = 150 for primary, 130 for secondary gears.

With hard on soft gears (pinion to wheel) K = 240 for primary, 210 for secondary.

With case hardened gears K = 360 for primary, 320 for secondary.

The above are maximum permissible 'K factors', in practice lower values would be adopted, but it can be seen that the tooth loading has increased four or more times since the earlier types manufactured.

Involute

An involute curve is a curve which would be generated by a fixed point on a taut cord which be being unwound from a cylinder (or drum) in a plane at right angles to the axis of the cylinder.

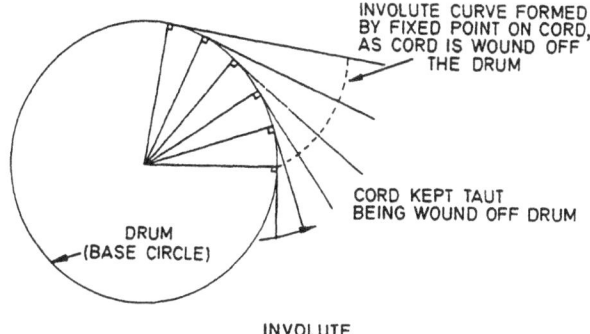

INVOLUTE

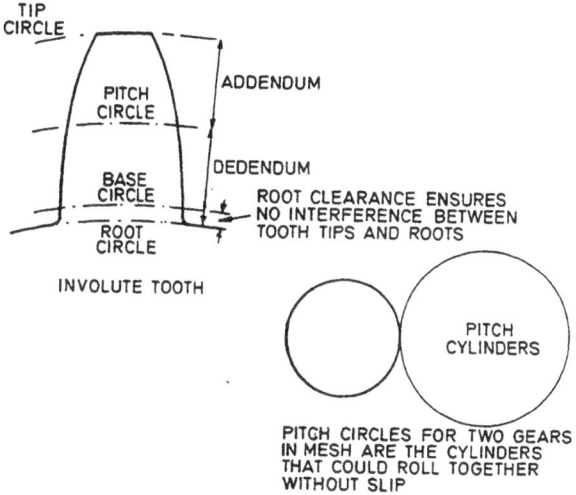

Fig. 3.1

The *base circle* for the involute is the section through the cylinder at right angles to the axis of the cylinder.

Fig. 3.1 shows in profile an involute tooth for spur gearing, additional depth is given to the tooth below the base circle to ensure that during operation no interference occurs between tooth tips and gear roots.

The reason why involute tooth forms are used is to ensure uniform speeds of rotation of driven gears, *i.e.* the velocity ratio of the two gears in mesh is kept constant.

Fig. 3.2 is an illustration of two spur gears with involute teeth, in mesh with one another. Assuming that the small gear is the driver and the larger the driven, a tangent PQ to the base circles can be drawn as shown. Where this tangent intersects the line of centres RS we have a point of special significance called the *pitch point*. This pitch point is also the point of contact between the pitch circles, and providing gears are designed so that the common normal at the point of contact between two teeth is along the line PQ through the pitch point, the gears will have a constant velocity ratio.

With the involute gearing as shown, contact between two teeth will begin at A and end at B. If we consider the point where the teeth are in contact, *i.e.* just before point B, and we draw a line normal to PQ through this point of contact then it will be a common tangent to the involutes of the two gears. Hence involute teeth fulfil the necessary condition for constant velocity ratio between gears.

PQ, along which tooth contact takes place, is called the *line of action* and the angle between the line of action and the common tangent to the pitch circles is called the *pressure angle*. Gear designers have standardised tooth pressure angles, they would be $14\frac{1}{2}$ or 20 degrees. The latter for high speed drives.

The normal force between gear teeth acts along the line of action and torque in the gearing would be the product of this force and its perpendicular distance from the gear centre (*i.e.* PS or RQ).

To avoid interference between gear teeth the tip circle radius of the driver must be less than RP and that of the driven wheel must be less than SQ, this ensures that the tip corner of one tooth does not cut into the root of another, remove part of the involute and hence reduce tooth strength.

Helical spur gearing is designed so that the helix angle ensures that one end of a helical tooth becomes meshed before the opposite end of a preceding tooth disengages. In this way more than one tooth is in mesh at the same time.

Advantages of helical spur gearing are:

1.   Greater distribution of load.

2.   Smooth operation—no sudden engaging and disengaging.

3.   Gears operate quietly.

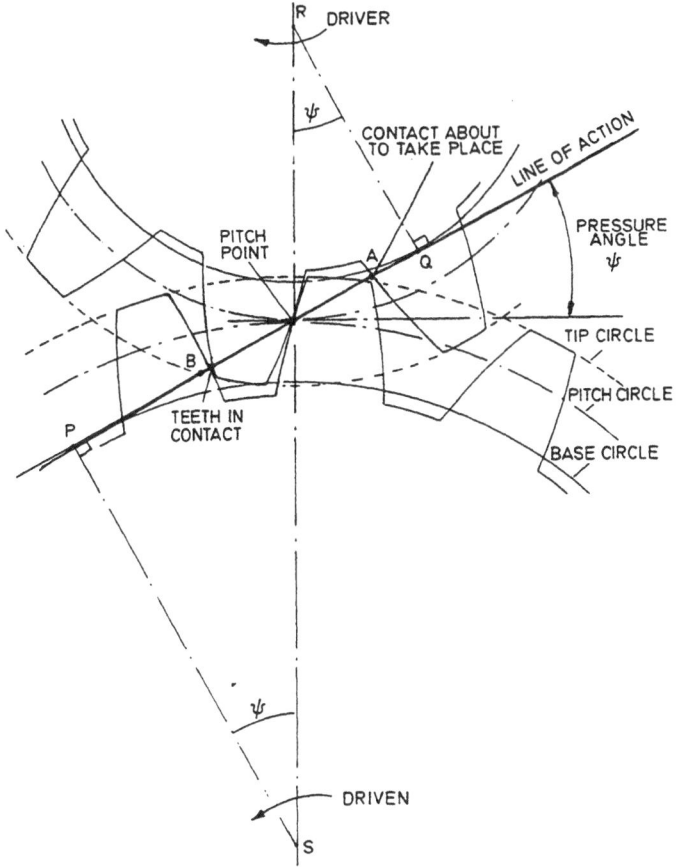

Fig. 3.2 INVOLUTE GEARING

With single helical gearing there will be an axial thrust in the shafts since the gear teeth are not parallel to the shaft. Double helical gearing overcomes this difficulty, the double helical gears have opposing helix angles, forming a herringbone pattern, which causes the axial thrusts to cancel each other out.

Helical gearing involute construction is shown diagrammatically in Fig. 3.3. The straight line AB on the plane, which is wound around the base cylinder, makes an angle with the base cylinder axis. This angle

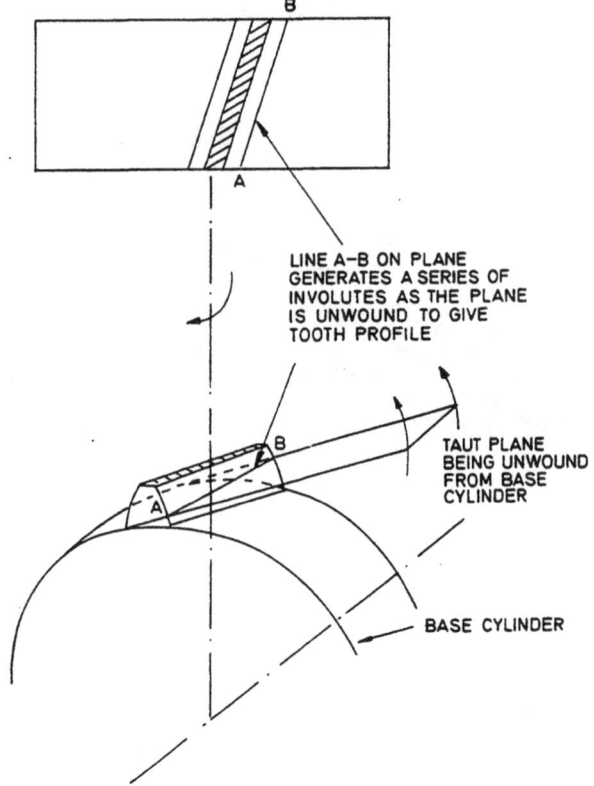

LINE A–B ON PLANE
GENERATES A SERIES OF
INVOLUTES AS THE PLANE
IS UNWOUND TO GIVE
TOOTH PROFILE

TAUT PLANE
BEING UNWOUND
FROM BASE
CYLINDER

BASE CYLINDER

Fig. 3.3 HELICAL GEARING (INVOLUTE CONSTRUCTION)

is the *helix angle* (or lead angle) for the gearing, since if the plane were unwound from the base cylinder its length would be πD (the circumference) and the lead of the tooth would be:

$$\pi D \times \text{Tangent of lead angle}$$

Helix or lead angles are about 30°.

EPICYCLIC GEARING

The main difference between parallel shaft gears and epicyclic is that in the former the wheel axes are fixed whereas in the epicyclic at least one axis moves relative to another fixed axis.

An epicyclic gear consists of a sun wheel on a central axis, a internally toothed ring called the annulus, planet or star wheel carrier and plant or star wheels which revolve on spindles which are attached to the carrier.

With the same basic arrangement various gear ratios are possible for the same sized gears if different items are used for input, output and fixing.

Planetary arrangement Fig. 3.4

In this case the annulus is fixed and the high speed shaft would be connected to the sun wheel, the lower speed output on the planet carrier shaft. This arrangement gives speed reductions of 3:1 to 12:1 depending upon relative gear wheel dimensions.

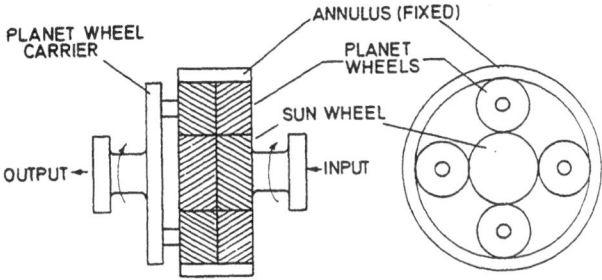

PLANETARY GEAR

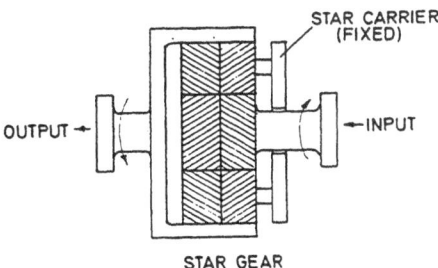

STAR GEAR

Fig. 3.4 EPICYCLIC GEARING

Star arrangement Fig. 3.4

Here the carrier is fixed, high speed input on sun wheel shaft and output annulus carrier shaft. Input and output revolve in opposite directions and with this arrangement speed reductions of from 2 : 1 to 11 : 1 depending upon relative gear wheel dimensions, are possible. Such a gear (strictly not an epicyclic since planet wheel axes are stationary, *i.e.* why the system is called star) would be useful for the first stage reduction in a h.p. turbine set as the centrifugal effect of planets would be reduced.

Solar arrangement

This is used for special application since the gear ratio is low about 1.1 : 1 to 1.7 : 1. The input goes to the annulus, the output from the planet wheel carrier, and the sun wheel is fixed.

GEARING ARRANGEMENTS

Typical gearing arrangements in use are shown in Figs. 3.5 and 3.6, one turbine drive is shown in each case, the pinions for multi-turbine drives are round the wheel periphery.

Single reduction gearing

Is mainly used for turbine driven alternators and pumps. For main propulsion the high turbine speeds needed for high pressure and high efficiency operation require reductions of over 50 : 1 and the limit for single reduction gearing is about 30:1 since a 150 mm diameter pinion would require a 4.5 m diameter wheel, which is too large.

A typical single reduction gear ratio is about 22 : 1 with turbine speed about 2000 rev/min. The term hunting tooth is applied to illustrate the fact that a non-exact gear ratio is usually used. For example a 30 tooth pinion at 22 : 1 would require a 660 tooth wheel for an exact ratio. In practice 659 or 661 teeth are utilised so permitting every tooth on the pinion to mesh in turn with every tooth on the wheel giving an even wear distribution.

Double reduction gearing is commonly used with turbines whose speeds go up to about 7000 rev/min but where large slow running propellers are used triple reduction gearing would be employed for turbine speeds of 6000 rev/min or more. A typical double reduction

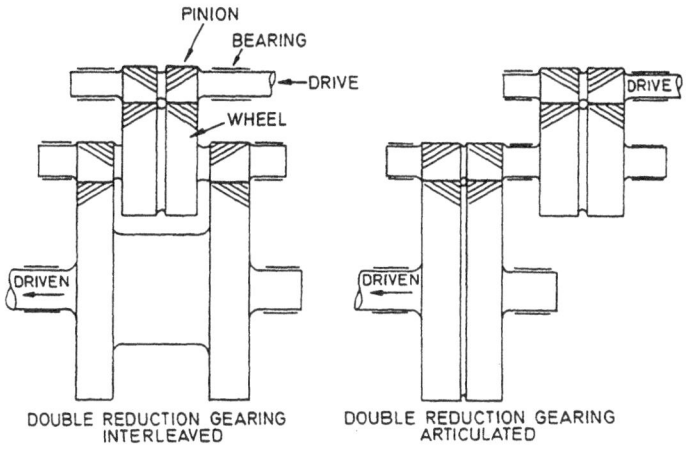

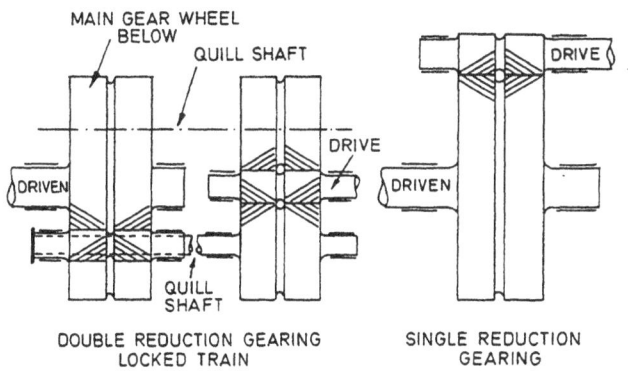

Fig. 3.5 TYPICAL GEARING ARRANGEMENTS

could be:—turbine speed 6600 rev/min, propeller speed 100 rev/min, 66:1.

The interleaved design is sketched, an alternative where the primary wheels straddle the secondary wheels is more compact and is generally referred to as nested.

Articulated gearing allows easier removal of individual trains and gives greater flexibility as regards alignment with less risk of faults of pitch being superimposed.

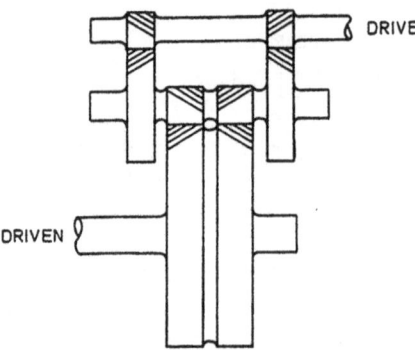

Fig. 3.6 DOUBLE REDUCTION GEARING INTERLEAVED (SPLIT PRIMARY)

The locked train design has been mainly employed in R.N. practice, sometimes called dual tandem gears, and is lighter and more compact. In this arrangement, power from the turbine is transmitted from the turbine to two primary reduction wheels by a centrally disposed single pinion. The loading on the wheels is thus halved as compared with the usual single primary wheel drive. The lower tooth pressure permits of a more compact unit. A further advantage is that the tangential thrust of the primary pinion is neutralised, thereby greatly reducing the pinion bearing loading. Each primary wheel is connected to its secondary pinion by means of a quill shaft and flexible coupling. Thus the drive absorbs any torque inequality between the two pinions, thereby equalising the load. This extremely compact type of reduction gear was used for all new British steam turbine driven warships and may be found in use for some high powered merchant ships. The sketch shows a small single turbine unit; h.p. and l.p. turbine installation will employ the same type of gear arrangement, except that the second reduction will use four pinions instead of two. On the sketch the primary and secondary sets have had to be drawn a distance apart for clarity. In practice they are very close and the quill drive is very short.

The use of a quill (nodal) shaft gives a reduction of weight, gives greater flexibility in absorbing misalignment and may also be utilised at the design stage to alter shaft stiffness due to the hollow 'tube' so amending vibration characteristics in various speed ranges. The fine tooth design of involute tooth is utilised, many of the features are similar to the flexible coupling (gear type) described later, Fig. 4.4. This quill shaft coupling is used in locked train and articulated gear sets in many cases. Gearing efficiency is about 98% at full power for most designs. A quill shaft intermediate speed coupling is as shown in Fig. 3.7.

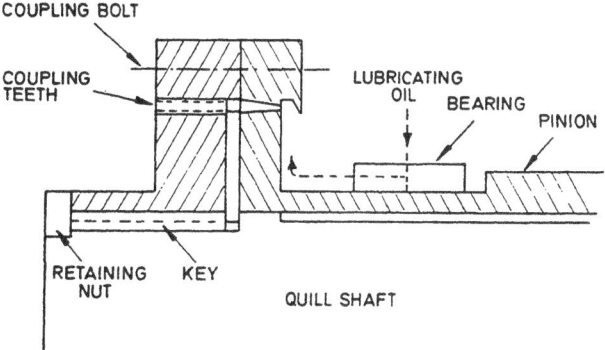

Fig. 3.7 QUILL SHAFT INTERMEDIATE SPEED COUPLING

A triple reduction gear as used in the Stal-Laval turbine system is shown diagrammatically in Fig. 3.8.

This system is used for slow running propellers (80 to 95 rev/min), large diameter slow running propellers have higher propulsive efficiency but some of this is offset against the slight reduction in gearing efficiency, the greater number of gearing reductions the smaller the gearing efficiency.

The h.p. turbine unit has a star gear for first reduction (as mentioned previously this reduces centrifugal effect), planetary for second reduction and parallel shaft pinion to bull gear for third reduction. l.p. turbine unit has double reduction as shown.

Rounded off typical speeds would be:

	Triple reduction h.p.	Double reduction l.p.
1st reduction	6000/1400 rev/min	4200/700 rev/min
2nd reduction	1400/ 400 rev/min	700/ 80 rev/min
3rd reduction	400/ 80 rev/min	

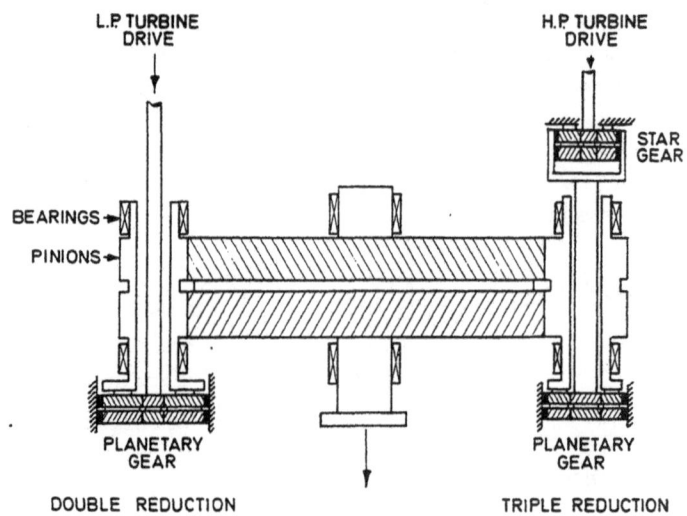

Fig. 3.8

GEARING CONSTRUCTION AND MATERIALS

Early designs utilised a heavy cast iron gear case. Gear wheel centres were usually of cast iron (with suitable lightening holes) fitted onto a shaft taper, keyed and secured by a large nut. Wheel rims were shrunk on and secured axially by screws or keys between rim and wheel centre.

Modern practice utilises an all welded prefabricated steel gear case, it is important that it be sufficiently rigid to avoid undue deflections under varying load conditions and maintain bearing alignment to ensure uniform tooth loading. Generally the gear is supported at two to four points in order to minimise the effect of ship strains due to cargo distribution and sea-ways.

The gear wheels consist of steel web-plates welded to gear wheel tyre and hub. A completely automatic electro-slag or submerged arc welding processes is used on the pre-heated tyre and webs (pre-heated about 250° C) the remainder of the gear welding is usually done by hand, After welding, the structure is stress relieved at 600° C and inspected by X-ray and magnetic crack detection processes. The steel gear wheel centre is keyed and butted against the shaft collar nut, there is no taper arrangement.

The wheel tyre, or rim, could be through hardened steel to British Standard Specification En16C, of composition:—Carbon 0.4% Silicon 0.3%, Manganese 1.5%, Molybdenum 0.3% u.t.s. 780 MN/m^2.

Pinions could be forged steel to British Standard Specification En 36C, of composition:—Carbon 0.15%, Silicon 0.2%, Manganese 0.6%, Nickel 3%, Chrome 1%, Molybdenum 0.15% (one of the 3% Nickel steels) u.t.s. 1000 MN/m^2, case hardened to a depth of about 2 mm.

Double helical pinion teeth are usually harder than wheel teeth since they are in use to a greater degree. It would be normal for them to be hobbed, case hardened, hardened and ground. Wheel teeth may be just hobbed.

During manufacture the teeth are normalised after rough machining to remove stresses, any hobbing or shaving must be carried out under controlled temperature conditions to avoid slight undulations in the teeth due to variation in expansion. Nitrided, induction hardened and flame hardened gears are also used, but it is essential that pinion and wheel teeth be bedded accurately to each other across the full tooth face.

With epicyclic gearing, sun and planet wheels would be similar in manufacture to pinions and generally nitrided. If the tooth loading factor is high (*i.e.* high K) the annulus would also be nitrided.

Gear cases

Design objectives may be as follows:

1. A simple inexpensive structure.
2. Compact, to save space and simplify transport problem from manufacturer to ship. Yet, having good access for inspection purposes.
3. A one piece fabricated structure.
4. It must be rigid externally to avoid twisting, yet it must be internally flexible to accomodate tooth forces.
5. A simple support arrangement to the ships structure to which the weight and variable reaction forces are transmitted.
6. To fit easily to standard hull frame-work.

7. Casing must be oiltight, hence it must be designed to have as few joints as possible.

Accomodating the main gear wheel can present problems. If the speed reduction is large, which is the modern trend for slow turning propellers, the main (or bull) gear in a double reduction arrangement could be large. This increased size and weight affects the gear case dimensions, supports and position in the vessel. Use of epicyclic gearing, increased primary reduction or triple reduction gearing can maintain the main wheel dimensions within tolerable limits.

A modern single piece gearcase for a cross compound turbine using double reduction dual tandem articulated gearing is shown in Fig. 3.9. The casing has no bottom half but a light, oil-retaining, trough is fitted. Main wheel bearing support is secured from below with vertical bolts, in addition horizontal bolts together with keys and dowels may be provided.

Gear teeth records

It is essential that records of gear teeth condition be kept. Means of obtaining the record may vary but it must be kept up to date for comparison purposes. Changes in readings recorded, however small,

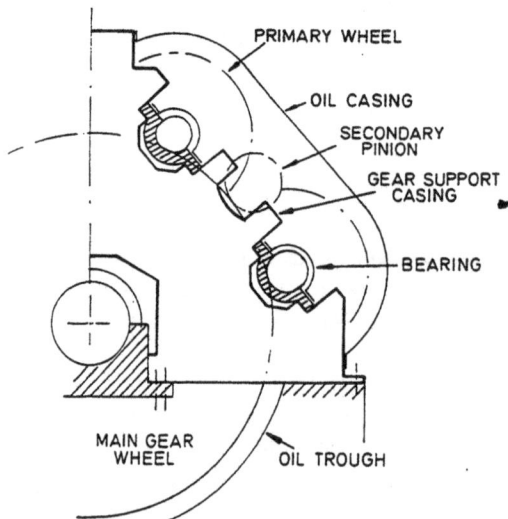

PRIMARY WHEEL

OIL CASING

SECONDARY PINION

GEAR SUPPORT CASING

BEARING

MAIN GEAR WHEEL

OIL TROUGH

Fig. 3.9 MODERN GEAR CASE

indicate to the operator all is not well. He should investigate the reason for the change.

A simple record may consist of sketches of numbered teeth (from some reference mark for identification) with a brief description of the degree and type of damage (as shown in Fig. 3.10). Alternatively, a type of 'brass rubbing' of the tooth surface may be taken using a soft lead pencil and paper. Or, marking blue may be lightly smeared across the tooth surface which is then wiped and a piece of transparent sticky plastic tape is placed over the tooth surface, this when rubbed over picks up the marking blue from the recesses. The tape is then removed and stuck into the record book alongside the tooth identification number.

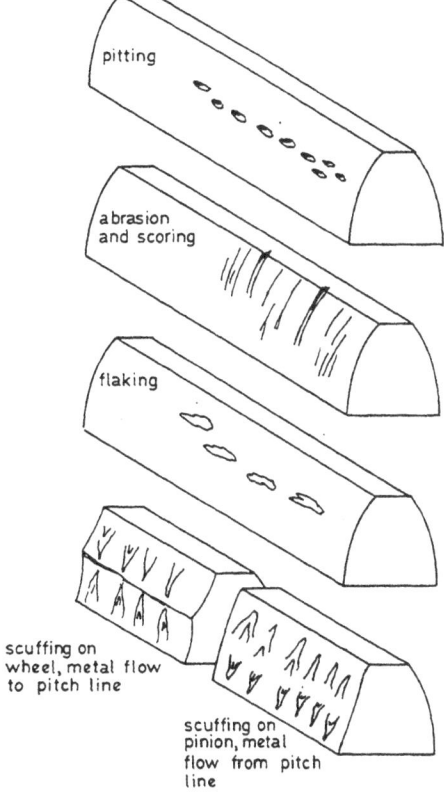

pitting

abrasion and scoring

flaking

scuffing on wheel, metal flow to pitch line

scuffing on pinion, metal flow from pitch line

Fig. 3.10

GEARING FAULTS

To reduce the risk of gearing faults it is essential that alignment is perfect, that the correct lubricant is used over a long running in period, and that the lubricating oil is subsequently kept in first class condition. Causes of faults, other than those implied above, could be one or a combination of the following; incorrect heat treatment, shock loading, vibration, bad surface finish and cutting.

Pitting

New gear teeth may have variations in surface smoothness and hardness, and the high and hard spots carry most of the initial loading under cyclic conditions. This local intense cyclic loading results in sub surface fatigue and fine fatigue cracks. Lubricating oil can enter these cracks and intense hydraulic pressure, caused by the closing of the teeth upon re-engagement, can lift out metal particles from the surface of the gear. The defect is generally more pronounced at or near the pitch line since this is the region of maximum load on the gear.

The characteristic rounded cavities or pits may link up leaving very little carrying areas, and eventually tooth fracture about the pitch line may occur.

Intense, isolated, relatively deep pitting may be found at the ends of the gear wheel teeth, this may be due to insufficient end relief, *i.e.* the chamfering or rounding off of the tooth profile to minimise the possibility of stress raisers.

Incipient pitting, sometimes called corrective pitting, may occur in new gears. After the hard or high spots break away the gear may wear more evenly until the surface is smooth and polished.

Scuffing or scoring

If the oil film between the teeth breaks down, metal to metal contact between surface asperities can occur resulting in local fusion or welding. This is caused by high temperature generated by friction as the surface asperities move across each other under high load. As sliding continues the welded metal is torn apart and the tooth working surface damaged.

It is normally found on the softer of the two gears in mesh and usually occurs during recession since the damage is generally found on the dedendum of the wheel and the addendum of the pinion.

There should be no difficulty in recognising defect due to scuffing, the surface will be dull and rough compared to the unaffected surface. The scuffed surface when examined with low magnification will be seen to be torn and scored in the direction of sliding, working from the root of the wheel tooth to the pitch line.

Interference wear

If the roots and tips of the gear teeth have not been carefully relieved it is possible for the tips to dig into the roots causing wear of the wheel roots (the softer material).

Abrasive wear

Solid particles in the lubricating oil causing scoring of the teeth. Grooving of the tooth surface in the direction of sliding and possibly some of the hard particles becoming embedded in the teeth are indications of abrasive wear.

Flaking

Usually confined to case hardened gears and could be caused by poor heat treatment combined with any condition which stresses the metal beyond its yield point. Flakes of hardened metal break away from the surface due to the high load.

Plastic flow

Local stresses tend to form a wave in the metal which rolls ahead of the point of contact, this results in sub surface fatigue failure and could result in ridging or even flakes of metal being sheared from the surface.

TEST EXAMPLES

Class 2

1. With reference to gearing explain the following: (a) spur gearing, (b) double helical gearing. Explain the advantages of helical gearing over straight spur gearing.

2. With reference to gearing explain the following:
(a) Addendum, (b) dedendum, (c) line of action, (d) pressure angle.

3. What is the shape of a modern gear tooth? How is lubrication

affected? What advantages and disadvantages are there with this form of construction?

4.   Sketch and describe the gearing used for a turbine having double reduction gearing. How is the main gear wheel constructed and what materials are used for the gears?

5.   Write explanatory notes on the following items which apply to marine gearing: (a) pitting, (b) abrasive wear, (c) tooth fracture.

Class 1

1.   Sketch and describe a nodal drive arrangement suitable for a turbine plant. Discuss the advantages to be gained by using such a drive.

2.   Describe, with the aid of sketches if required, the various types of defect that can occur in marine gearing.

3.   Discuss the materials from which pinion and gear wheels are manufactured and sketch any form of gear train with which you are familiar, naming the type to which you refer, and the speeds of each item in the gear train.

4.   Sketch and describe the following gear trains: locked train, nested, articulated.

5.   Some turbine installations employ triple reduction gear drive. With the aid of a sketch explain how this is achieved and explain the difference between epicyclic gears of the star and planetary type.

6.   Explain the following defects which occur to gearing: pitting, abrasive wear, spalling or flaking, initial pitting or incipient pitting.

CHAPTER 4

TURBINE PRACTICE

In general a whole volume, or volumes, could be given under the above heading. This chapter will attempt to condense the subject into the basic essentials. In this respect the basis used will be the questions set in the subject at the D.O.T. examinations. The work will be divided up into four main sections:—Turbines in general, Impulse-Turbine details, Impulse-Reaction Turbine details, Associated Equipment details.

TURBINES IN GENERAL

Warming Through Procedure

This practice varies greatly as two of the main factors are: 1. The age and design of the turbines and 2. Type and duty of ship. For example, three methods in outline, among many are:

1. A six hour period, almost full vacuum, manoeuvring valves cracked open, gland steam on, circulating water on, lubricating oil on, no use of turning gear, turning on steam for a few revolutions after about 4 hours for $\frac{1}{4}$ hour intervals, etc.
2. Warming up from cold, boilers and engines, all systems and valves open, gradual shutting in, fairly continuous use of turning gear, etc.
3. A 2 hour period, low vacuum, warming through valve open, gland steam on, circulating water on, lubricating oil on, continuous use of turning gear, etc.

The first method is older practice, the second method was utilised in R.N. practice, the last method is more modern practice and would be a preferred method. From the *examination* view-point the best approach for a student is to describe the system utilised *in his own experience*.

To give some information, for comparison, the following procedures are given for warming through turbine installations.

The description given for Method 1 is for impulse-reaction turbines of Parsons type based on a ship of about 15 years of age. Included in this are also details of manoeuvring and general shut down and emergency stop procedures.

The description given for Method 2 is for turbines in general based on slightly more modern practice although it is by no means the latest or standard practice. Also included are some general remarks relating to distortion, etc.

Method 1

1. All drains are opened and the l.p. exhaust temperature should never exceed 105° C. A *slow* warm through is practised of about 4 hours duration (minimum).

2. Main circulating, extraction and lubricating oil pumps are put *on* (appropriate valves, *e.g.* oil run down valves, are previously checked as *open*). Adjustable thrust blocks (if fitted) are set to maximum clearance, *i.e.* 'Contacts off'. Engines are turned with turning gear then turning gear is removed. Manoeuvring valves and guard valves are 'cracked' open, gland steam is put on, air ejector is put on to give about 0.85 bar, test of emergency stop valve is carried out, expansion of sliding feet are checked for no obstructions and the initial reading noted on expansion gauges. Restriction of lubricating supply is often applied to allow the oil temperature to reach 32° C quickly, it is then fully opened.

3. Vacuum is raised to about 0.5 bar after $2\frac{1}{2}$ hours and engines are turned slowly ahead (and astern) on main steam for a few revolutions only about every 10 minutes for the next $1\frac{1}{2}$ hours. (Test briefly for full vacuum $\frac{1}{2}$ hour before stand by.)

Note.

A stop should be provided at the manoeuvring valves so that if the engines cannot be turned the valves can be set to allow more warming through steam after about $1\frac{1}{2}$ hours from the start, for about 1 hour, and then the turbines can be allowed to 'soak' until a turn on main steam can be arranged.

4. During manoeuvring the drains are regulated as required. At long stops vacuum will be dropped to about 0.5 bar.

5. At 'full away' drains are mainly shut, astern guard valves are closed, 'Contacts On' (if end tightened blading), bled steam valves opened, turbine clearance gauges are checked, etc. The routine adopted depends very much on the individual ship and auxiliary plant.

6. For entering port the routine given in Section 5 is virtually reversed. At shut down for extended stays in port, with engines not required, the main circulating pump and lubricating oil pump are normally run for about 1½ hours and then shut off. All valves are then shut, including the emergency stop valve.

7. The correct temperature gradient is vital during warming through periods. Long stand by periods are also particularly difficult if the engines cannot be turned. Distortion is very liable to happen at these times.

8. In the event of an emergency stop of the main engines at sea then the manoeuvring valves should be instantly shut. The astern guard valve is then opened and astern braking steam can be applied. It is best to move to 'Contacts Off' if the turbines have end tightened blading before applying astern steam. The emergency stop valve is normally arranged to close so as to mask ahead steam only so that astern braking steam is available to bring the engines rapidly to a stop.

Method 2

The time period utilised (for turbines) is about 1½ hours. Distortion results from variations of temperature gradient which is most liable to occur during warming through or stand by periods.

The main causes of distortion are:

(a)   Stop periods at full vacuum.

(b)   Local overheating by use of gland steam only.

(c)   Local overheating by steam admission of 1. Live steam through restricted nozzle groups or auxiliary connections. 2. Exhaust steam to main condenser or 3. Direct live steam to a cold astern turbine.

To avoid distortion the turbines ideally should be uniformly heated to *the running temperature gradient.* Local overheating must be avoided and excessive condensation prevented, adequate drainage is essential. Turbine rotation during the danger periods is almost essential and arrangements between deck and engine departments to allow for this must be *clearly established and understood.*

1. All main valves between boilers and h.p. turbine, up to and including the final warming through valve, should be opened and other valves, e.g. gland steam, auxiliary heating, manoeuvring valves, etc should be *shut*. The boilers are then flashed away, all drains on valves, piping and turbines should be fully open. The manoeuvring valve should have a lock and if this is fitted the turning gear can be operated regularly or continuously.

2. The heating rate should be controlled so that inlet temperature to the l.p. turbine reaches about 75° C after *one hour*, if the turning gear is in action it can now be removed. The vacuum should then be raised as rapidly as possible, by putting gland steam on, to about 0.5 bar to 0.34 bar and the turbines turned a few revolutions on steam. Only ahead rotation is normally preferred, depending on the arrangement, unless it is essential to use astern steam to avoid mooring difficulties, astern steam can cause severe distortion to a cold, fairly static, astern turbine. In this respect much depends on the design of the warming through valve arrangement as well as the turbine set, astern turbine warming through arrangements by suitable valves could well be provided with advantage to obviate such difficulties. The temperature of the exhaust to condenser space should be kept under continual observation. Vacuum should *not* exceed 0.34 bar, valves can be shut in or regulated as steam pressure rises. If main boiler steam is immediately available then the above procedure is applicable, provided proper warming and draining of pipes and fittings is carried out.

3. The warming through valve is now *closed* and ahead manoeuvring valve *opened* slightly. The turbines should then be turned under steam at 2 minute intervals for 15 minutes. The vacuum should now be fully raised to test the system and then dropped to 0.34 bar. Engines are now ready for stand by.

4. If turbines are so prepared, they should, if *not* being used, be moved under steam at intervals not exceeding 10 minutes. Grave risk of damage can result from maintaining turbines in a state of immediate readiness, without rotation under steam at short intervals of time. If it is essential that rotation can not be carried out then all steam to turbines should be shut off and gland steam also shut off. If rapid starting is however still required the gland steam can be left on but at a minimum to keep the lowest reasonable vacuum for an efficient start. If possible after such conditions the rotation at 2 minute

intervals for 15 minutes, at 0.34 bar vacuum should be repeated before bringing turbines into normal operation.

5. Drainage is vital to avoid distortion. All drain valves must be clear and should not be closed until the turbine has been rotating for several minutes, they should be opened at slow speeds or stop periods. The possibility of leakage past valves and subsequent water accumulation must always be allowed for.

6. Auxiliary exhaust steam entering main condensers should be reduced to a minimum at stand by periods. Partial cooling of turbines after steaming results in distortion and should not be allowed. If at all possible engines should be slowly moved ahead and the vacuum reduced. Steam should not be admitted to turbines at full vacuum unless they are allowed to rotate.

7. During prolonged astern running the l.p. turbine temperatures and expansion indicators must be closely watched. L.p. astern casing temperatures *should not exceed* 250° C and exhaust spaces *should not exceed* 120° C.

8. In the event of unusual noises from turbine machinery the turbines must be stopped and allowed to stand for at least 15 minutes without steam before attempting rotation again. All types of clearance indicator should be closely watched at all times.

Turbine materials

The following is intended as a guide. Trade names of materials and detailed material composition have deliberately been avoided. It should be noted that most examination questions on turbine details require the material used to be named and a brief outline of its properties given.

h.p. casings

3% molybdenum cast steel *or* 0.5% molybdenum, 0.3% vanadium cast steel.

l.p. casings

Cast steel. In certain cases fabricated mild steel.

Rotors (solid, hollow, built up, gashed)

Forged chrome-molybdenum steel or 0.5% molybdenum steel.

Blading

0.12% carbon, 12% chrome, 1% nickel stainless iron *or* 0.24% carbon, 12% chrome stainless steel *or* 0.25% carbon, 12% chrome, 36% nickel steel.

Nozzles

Stainless iron *or* 67% nickel, 28% copper monel metal.

Glands

Cupro-nickel *or* leaded nickel bronze (65% copper).

Packers and spacers

Rarely used. If fitted, brass or soft iron.

Turbine wheels

As for rotor material.

Diaphragm centres

Mild steel or 0.5% molybdenum steel.

Note.

Rotors should be normalised at about 900° C and annealed at 660° C for about 48 hours. All casings should be annealed. Mild steels are often used to 427° C. Austenitic stainless steels (18% chrome, 8% nickel, 1% colombium) have been used to 650° C. Nimonic steels which have a very high nickel and chrome content with additions such as titanium have been used for still higher working temperatures.

Creep, strength at high temperatures, erosion resistance and working temperature are the main governing factors for material choice. Some of the elements mentioned in the foregoing list are important because they improve in some way the properties of the material, *e.g.*

Nickel, increases strength and erosion resistance.
Chromium, increases resistance to corrosion and erosion.
Molybdenum, increases strength at high temperatures.
Vanadium, increases strength and fatigue resistance.
Practically all turbine rotor forgings are made today of vacuum

treated steel. This process reduces the dissolved Hydrogen level and minimises the risk of thermal flake formation as forging cools after pressing.

Advantages and disadvantages of turbines

Consider the following advantages of turbines:

1. Uniform torque under steady load.
2. Few contact friction parts.
3. No internal lubrication.
4. Low centre of gravity in ship.
5. High power weight ratio in a small space.
6. Good static and dynamic balance.
7. Reduced maintenance.

Consider the following disadvantages of turbines:

1. High speeds require reduction gearing for propulsion.
2. Not directly reversible, astern turbine required.
3. Low starting power.
4. Manoeuvring can be slightly sluggish.

Consider the following on fuel consumption:

The specific consumption (all purposes) of an internal combustion reciprocating engine is about 0.23 kg/kWh. Specific fuel consumption for turbines has reduced in the last twenty years from about 0.37 to 0.24 kg/kWh so that the modern turbine is highly competitive. To achieve such low fuel consumption rates however it is essential to have high efficiency boilers, employ very high steam temperatures, utilise reheat, utilise auxiliary drives from main engine, etc., as well as designing the turbine itself for maximum efficiency.

Double casing turbines

Use of high temperatures and pressures introduces a difficult design problem. The casing has to be made thicker to resist higher pressures and also act as the main support girder for the turbine rotor itself. The expansion on such a structure is difficult to allow for whilst maintaining correct axial clearances along the turbine rotor length. One solution is to introduce a double casing. The inner casing is pressure strength resistant but is of simple design and free to expand with ability to absorb thermal stresses arising with the large expansion due to the use of elevated steam temperatures. This inner casing does not however have to act as the main strength and support girder. The inner casing is fully supported by the outer casing which acts as a support cradle. The outer casing can be made as a rigid support as it is only subject to exhaust steam pressure and temperature. This casing maintains concentricity and alignment irrespective of expansions.

Exhaust steam between the casings acts as a steam jacket and reduces heat losses from the surfaces which improves thermal efficiency.

Referring to Fig. 4.1:

The cast steel inner cylinder barrel contains the diaphragms and is

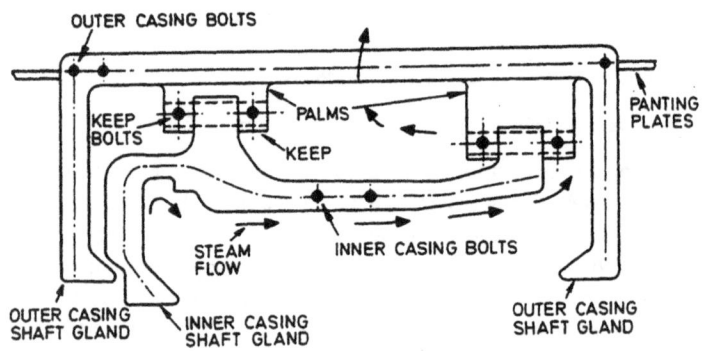

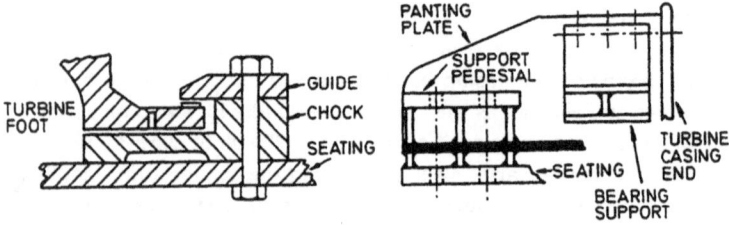

Fig. 4.1 EXPANSION ARRANGEMENTS

supported from and within the fabricated outer barrel support structure. Locating keys, in a fore and aft direction, in the lower half casing are usually fitted and in the sketch shown the inner cylinder is supported and secured by keeps and bolts to the palms of the outer structure. Fore and aft expansion allowance together with accurate location are desirable. All steam connections to the inner cylinder are given radial expansion allowance through the outer casing.

It should be noted that allowance for casing expansion (outer) is by panting plates whose flexure can adequately deal with such expansions. This practice has largely superseded the use of the sliding foot type of expansion and seating arrangement.

The turbine sliding foot arrangement is, however, also sketched for reference. The principle should be clear from Fig. 4.1. Note that expansion of casings is normally allowed in one direction only, that is, away from the fixed attachment forward from the gear box.

Fig. 4.2 shows diagrammatically support and expansion arrangements for a high pressure turbine. The arrangement allows the casing to expand axially, from the gear box end, and radially. Whilst maintaining accurate location at all times.

Two sliding feet supports below casing centre line have axial keys for location as shown—these must be kept free by means of high temperature lubricant. Four support palms, two aft which do not permit axial movement and two forward which do, are connected to horizontal extensions of the casing joint complete the arrangement.

Rotor position relative to the casing is controlled by the thrust bearing at the forward end. Internal rings, correctly dimensioned and fitted in the thrust bearing, should ensure no further adjustment is necessary.

Careful study of the turbine basis plans on subsequent pages will show different expansion arrangements e.g. vertical location keys, elongated holes etc., but all turbine casings and rotors must be allowed to expand freely whilst maintaining accurate location to seatings and to each other. Otherwise distortion, stresses and damage due to rubbing etc. can occur.

Piping connected to the casing must be flexible to avoid restraining or causing casing movement which could lead to stresses, misalignment and possible damage. Bellows pieces or sleeve expansion connexions for large diameter piping and large radius bends or pipe within a pipe system for small diameter.

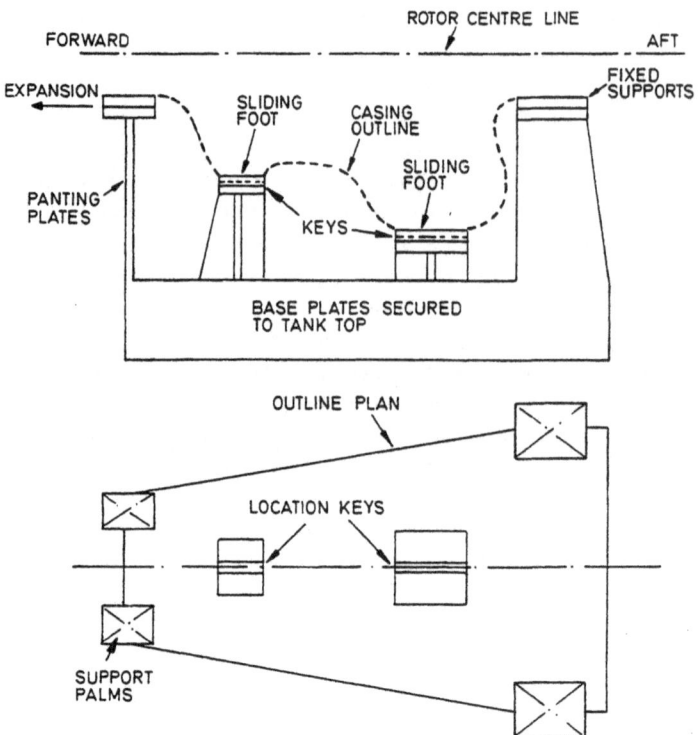

Fig. 4.2 MODERN TURBINE SUPPORT AND EXPANSION ARRANGEMENT

Turbine clearance gauges

The operation and application of such gauges should be clear from the sketches. The axial position of the rotor is of vital importance when considering impulse-reaction turbines with end tightened blading due to the close axial clearances used, this is of less importance for impulse turbines.

L.p. turbine water extraction

Moisture in low pressure steam can cause water braking with lowering of turbine efficiency and also erosion of the leading edges of blades. The maximum amount of moisture is radially thrown out from the trailing edge of the moving blade. Three drainage belts may reduce moisture content by as much as 30%.

Flexible couplings

The flexible coupling is used to prevent slight axial movements,

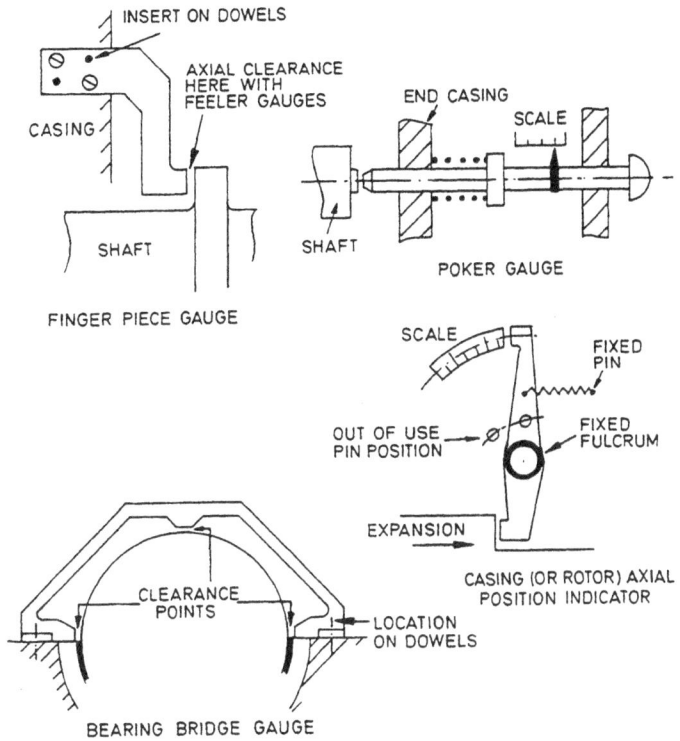

Fig. 4.3 SOME TYPICAL TURBINE CLEARANCE GAUGES

mainly due to temperature changes, from being transmitted to the pinion as the wheel and pinion must run true relative to each other, without end thrust. The coupling also allows for slight changes in alignment. Slight relative movement between sleeve and claw does occur so that lubrication is essential. The lubrication supply from the adjacent bearing can be seen on the sketch, oil holes in the sleeve allow the oil to drain away. Radial clearance is about 0.375 mm. In many types the claw on both pinion and turbine shaft is a separate forging. In this case the forging is fitted to the shaft with a taper and double keys and secured at the shaft ends by a large nut. In Fig. 4.4 the claw is a separate forging fitted on the shaft directly. The main purpose of such a coupling is to allow axial flexibility between the turbine and the gear shafts, so permitting the turbine rotor to be held in

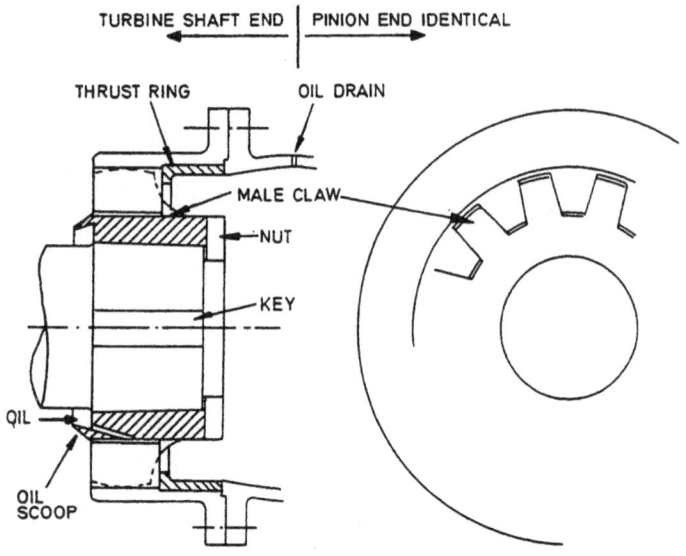

TURBINE SHAFT END | PINION END IDENTICAL

THRUST RING   OIL DRAIN

MALE CLAW

NUT

KEY

QIL

OIL SCOOP

Fig. 4.4 FLEXIBLE COUPLING (CLAW TYPE)

the correct axial position relative to the casing by the turbine thrust bearing, while the pinion can take up its correct running position relative to the gear wheel. Also it permits a slight degree of misalignment between driving and driven shafts, caused by bearing wear or due to change in the position of bearing pedestals.

Couplings are usually of the claw or gear type, the latter now being more widely used. Claw types tend to wear and form ridges or shoulders which prevent free movement, but this trouble does not exist to the same extent in gear types. In recent years in claw couplings trouble has been experienced with pitting and fretting corrosion.

Pitting may be caused by electrolytic corrosion rather than by stress fatigue. If stray electric currents (from whatever source) are passing along either shaft, intermittent metal to metal contact between mating teeth would allow current to pass through to the other shaft. Intermittent flow could result in local electrolytic corrosion.

Fretting corrosion as caused by vibration is difficult to overcome, but is less likely with gear couplings due to the greater number of teeth in contact. If a perfect oil film existed between the mating teeth there would be little risk of either form of trouble. Due to the very limited movement between the driving and driven members of the couplings, boundary lubrication exists. An extreme pressure oil might be beneficial by greatly reducing metal to metal contact.

It is usual to make the teeth on the female sleeve nickel steel whilst those on the male are forged carbon steel. Using dissimilar metals minimises possibility of fretting under high contact pressure.

When transmitting torque it could be possible for the couplings to lock and become inflexible. This could produce bending and un-balance with subsequent damage to pinions etc.

With claw types oil is fed to the claw teeth from the ends of the nearby rotor and pinion shaft bearings, and it should be constant in flow as well as in ample quantity to ensure free movement of the coupling. Drain holes provided in the sleeve permit oil to flow through (at least four holes of ample diameter). Due to high speeds of rotation there is a centrifuging effect and this may tend to choke the drain holes if there is any sludge present. Gear type couplings consist of two sleeves, or rings, having internal teeth, keyed to the rotor and pinion shafts respectively, with a distance piece or floating member having external teeth. The distance piece teeth are free to slide between the mating teeth in the sleeves, providing flexibility. The teeth are lubricated from the main system by means of small pipes feeding oil into the ends of the couplings, and then by centrifugal force to the gear teeth. After lubricating the teeth the oil escapes from the inner ends and then drains back into the oil tank.

It is possible due to centrifuging for the oil supply holes to the teeth to become choked with sludge—it must be remembered that speeds of the order of 7000 rev/min are possible. If this occurred, scuffing and excessive wear of the coupling teeth could take place. Excessive noise should give some warning that all is not right with the coupling.

Fig. 4.4 shows a detail of the claw type of coupling which has been in use for many years. The gear (or fine tooth) type of coupling is also illustrated in Fig. 4.5. Tips of external teeth on the distance piece of this form of coupling are machined to a spherical surface which allows good centring whilst the tooth profile, being barrel shaped, allows slight angular movement.

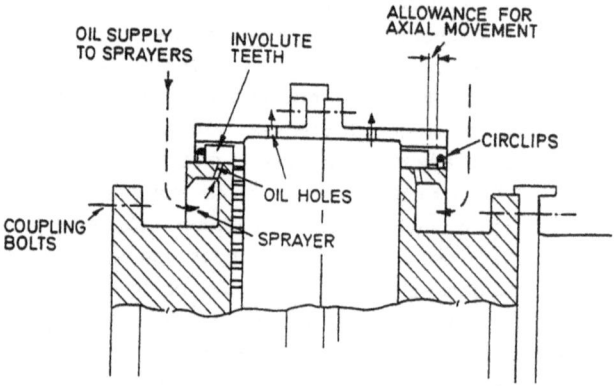

Fig. 4.5 FLEXIBLE COUPLING (GEAR TYPE)

Membrane type flexible coupling

Increasing use is being made of this type of coupling since they are: simple in construction, reliable, less expensive than toothed types, easy to assemble and require no lubrication. Fig. 4.6 shows a slightly simplified diagram of the type fitted to some of the more recent G.E.C. turbine installations. The unit allows for considerable axial movement and gives reduced gear tooth loading for equal transverse misalignment compared to the tooth type.

Thrust block bearings

A typical turbine thrust block has pads, retaining ring, liners, etc. The construction is virtually the same, except for size, as a standard main shaft thrust block (see Vol. 8) also the pads go fully round the shaft circumference and the lubrication is pressure feed. For examination purposes the sketch given later in fig. 4.28 of the adjustable thrust block is the usual requirement.

Turbine bearings

A typical turbine bearing is shown in Fig. 4.7, the shell for these high speed bearings may be either gun metal or steel dovetailed to receive white metal (tin 85%, copper 7% and antimony 8%) after degreasing, fluxing and tinning. The whitemetal layer, 0.5 mm thick gives reasonable margin for dirt absorption and slight mis-alignment,

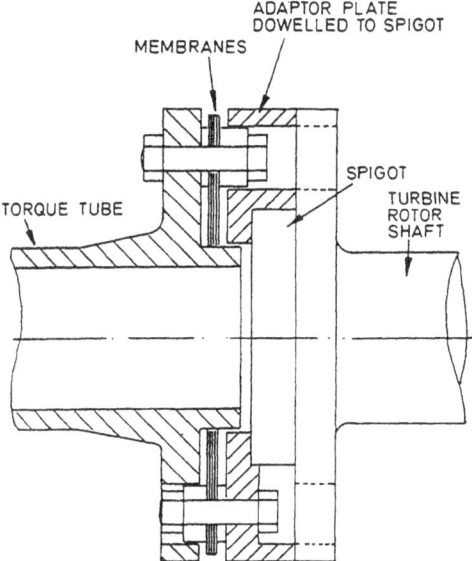

Fig. 4.6 MEMBRANE TYPE FLEXIBLE COUPLING

and in the event of oil failure the rotor shaft would run on the safety strip—bronze insert in the case of the steel shell. Spherical shells have also been used to assist in alignment.

Bearings must be symmetrical about a vertical plane through the shaft axis for running the turbine in either direction, and accurately located in the housing. Locking screws prevent rotation of the bearing in its housing, no liners are fitted, no oil grooves are cut in the white metal and the bearings can be removed without disturbing the shaft—the shaft would be supported on dummy bearings.

For machining after re-metalling reference diameters are provided at each end of the shell. When the rotor rests on the bottom of the bearing it is concentric with the turbine casing, and since the bearing clearance is 0.25 mm its axis must be eccentric with the casing. The reference diameters have the bearing axis as centre.

Bearing clearance is of extreme importance, if it is too small overheating of the oil can occur, and if too large misalignment.

Thin shell, or thin walled bearings as they are sometimes called,

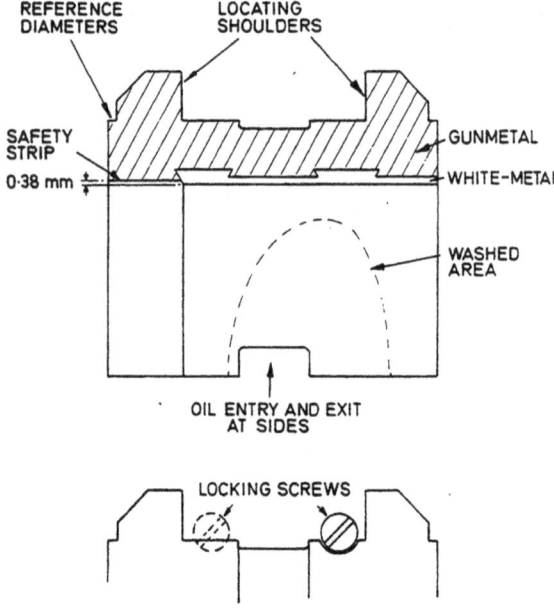

Fig. 4.7 TURBINE BEARING

rely on interference fit in their housing, hence the housing must be accurately machined to avoid the slightest misalignment.

Thin shell steel bearings (about 6.35 mm thick for a 230 mm diameter shaft) are a design which it is claimed will allow operation for an appreciable time under oil failure conditions. The thin steel shell has a sintered bronze backing thinly coated with white metal.

The white metal thickness is insufficient to cause blockage of oil ports of gland fouling. On restoration of oil supply the bearing will give good performance until the spare can be fitted. The optimum length-diameter ratio is between $\frac{1}{3}$ and $\frac{2}{3}$ dependent on duty. Turbine bearings invariably operate with full fluid film surface separation with little side leakage. The maximum bearing temperature at a given spot on the bearing does not always occur at the maximum load line. Chamfered bearings have the ability to accept a much greater thermal loading but with greater friction losses as the extra oil flow causes the bearings to run relatively cold.

Turbine glands

Glands are always of the labyrinth type in which the steam is throttled in passage through a small clearance existing between the edges of sharp metallic packing strips of brass, cupro-nickel or leaded bronze, and the rotating shaft or wheel hub.

In passage through the clearance the pressure drops and the velocity increases, the velocity is then dissipated in eddy motion in the space between the strips. Hence the reduction of kinetic energy will be dependent upon the number of gland rings and the clearance utilised.

Rubbing contact has occurred between gland strip and shaft which has caused local surface heating. This has in turn worsened the rub condition resulting in shaft bending or bowing. Invariably this condition originates due to eccentricity of the rotor caused by uneven heating, usually during manoeuvring conditions. Thermal straightening is possible but it is an expensive and skilled process. Packing strips should therefore be as thin as possible at tips and modern practice is to utilise spring backing on all gland strips.

Consider first internal glands as used in impulse turbines to prevent interstage leakage. The packing strips often in four segments are usually caulked in and are preferably fitted to a removal ring (in halves) which can be removed for replacements and repair machining without removing the diaphragm (which is also in halves). These sketches in Figs. 4.8, 4.9 and 4.10 are not all drawn to the same scale exactly but the dimensions given in Figs. 4.8 and 4.10 give a general idea of the sizes utilised in gland construction. The sketches in Fig. 4.9 illustrate two types of internal diaphragm gland for modern impulse turbines.

One design has four radial springs to each of four gland segments, *i.e.* two segments in each half of the diaphragm. The lower two segments are dowel pinned to the diaphragm to locate and prevent rotation.

The other design incorporates two leaf (or plate) springs arranged side by side to each of four segments. If contact is made between the gland packing (which is made of softer material than the rotor), the gland will heat up and move away from the rotor minimising damage.

Consider now the external glands used on all turbines for the shaft glands with steam sealing. The glands consist of a sleeve (in halves) which carries the labyrinth packing strips, the sleeve may be one

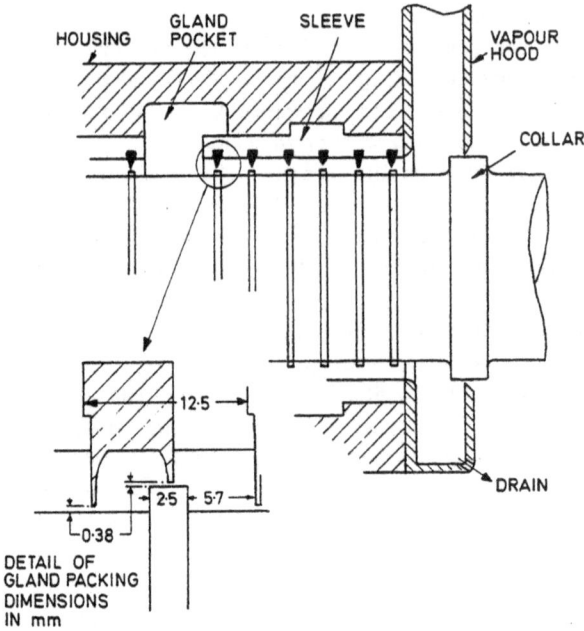

Fig. 4.8 EXTERNAL TURBINE GLAND

length carrying about 15 packing rings for h.p. glands or it may be in up to three lengths (h.p.) in series, again with a *total* of about 15 packing rings.

The profiles of modern packing rings are in general the *same* as described for the internal glands. Modern spring backed designs are also similar with the strip lengths spring backed into the gland sleeve instead of the diaphragm as previously sketched and described.

Alternative forms of external gland mainly used in older turbine practice will now be described. These are: 1. As fitted to impulse-reaction turbines for shaft gland and dummy piston gland seal (low pressure turbine end). 2. As fitted to impulse-reaction turbines for dummy piston gland seal (high pressure turbine end) when end tightening blading is used, this is a facial (axial) form of seal instead of radial. 3. Carbon ring glands which are not now in frequent use due to high shaft speeds and high steam temperatures used in modern turbines.

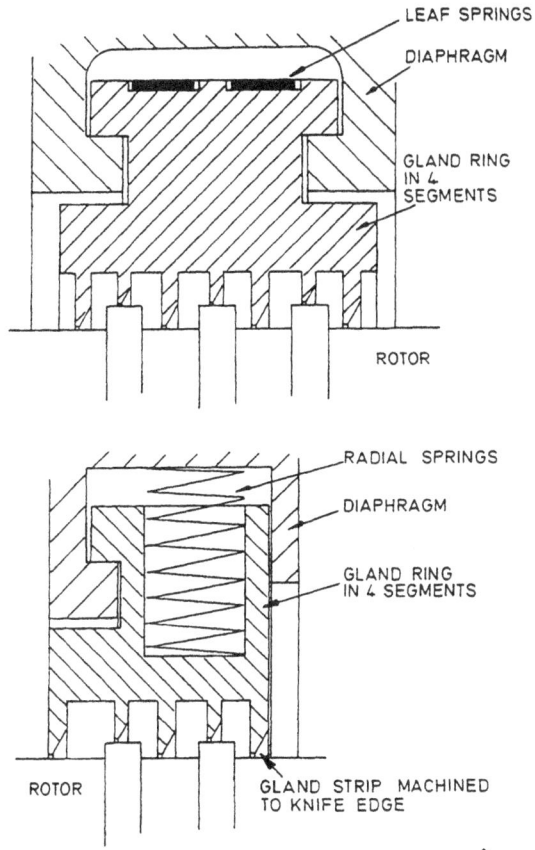

Fig. 4.9 INTERNAL TURBINE GLANDS

Referring to Fig. 4.10:

The first sketch illustrates gland strip design used in Parson's type turbines for many years. This radial clearance allows axial movement without affecting the clearance. The rows may be split into stages with pockets between, for h.p. two pockets, for lower pressures one pocket. This type is used for shaft glands. It is also used for dummies at the low pressure end of the cylinder as the dummy at the high

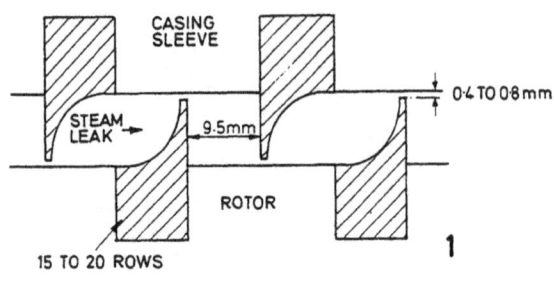

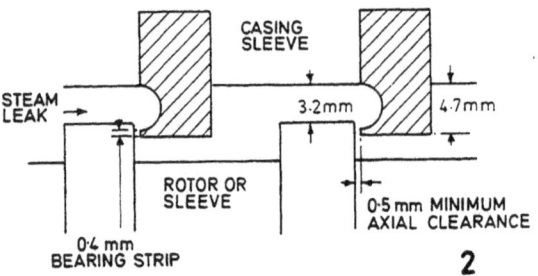

Fig. 4.10 EXTERNAL TURBINE GLANDS

pressure end is of the facial type with close axial clearance so that the other end of the turbine at the dummy must be unrestricted as regards expansion allowance.

The second sketch illustrates a facial or contact type mainly utilised at the high pressure dummy piston when end tightened blading is used. There are about 15–20 rows and the minimum axial clearance is usually near the blade minimum axial clearance (see later for dummy pistons and clearances in impulse-reaction type turbines).

The third type is the carbon block. The carbon rings are in three or four segments, held together by garter springs or leaf springs, a plate spring in the lower half takes the weight, rotation is prevented by dowel pins. The end clearance in the housing is about 2.5 mm and the radial clearance (when the shaft is at running temperature) is about 2.5 mm. The carbon rings do not require lubrication but it is essential that they are free to move in the casing housing sleeve.

IMPULSE TURBINES

General modern practice utilises such turbines for the high pressures and temperatures required to give maximum efficiency. A modern h.p. turbine will almost certainly be an impulse type whilst the l.p. turbine will be either impulse, impulse-reaction, or mixed, although the all impulse design is probably most favoured.

Steam conditions of 42 bar, 450° C, with a two casing cross compound set for powers of 10 500–15 000 kW (32 000–68 000-tonne tankers) are common practice. Such turbine sets give fuel consumptions near 0.3 kg/kWh and approaching 0.24 kg/kWh (for very large powers with efficient feed cycles and added refinements such as reheat). It seems probable that the single cylinder turbine, that became topical in the nineteen sixties because of its simple design and lower cost, will generally be limited to about 15 000 kW maximum power. At higher powers than this the cross compound set becomes increasingly more efficient, in fact at 15 000 kW the difference is about 1.5%. Some single cylinder turbines developing up to 26 100 kW have been built at an initial cost 15% less than an equivalent cross compound set, but due to the smaller number of stages its efficiency is about 3% less.

Hence for powers in excess of 15 000 kW and possibly even 7500 kW the two cylinder all impulse cross compound with gradually increasing steam pressures and temperatures to give corresponding reductions in specific fuel consumption seems to be current and foreseeable future practice.

In projected boiler installations pressures up to 105 bar and steam temperature 540° C may be employed to improve overall thermal efficiency, but maximum power per shaft for a cross compound set at present seems to be limited to 44 800 kW due to hull design considerations.

Turbine details considered in this section will be based on the above types. Basis plans for typical turbine sets are now given for *reference*. Generally, upper half of diagram shows sectional plan, lower half sectional elevation.

Single cylinder turbine

The design allows a simple, low cost and economical unit with a good balance of efficiency of both high pressure and low pressure sections.

Fig. 4.11 shows a cutaway view of 15 000 kW single cylinder impulse turbine propulsion machinery. With double reduction articulated gearing, two pass axial condenser, turning gear and inlets for main and overload steam. This unit has been installed on twin screw container vessels with steam supplied at 63.5 bar and 510° C from two Foster Wheeler ESD 111 boilers.

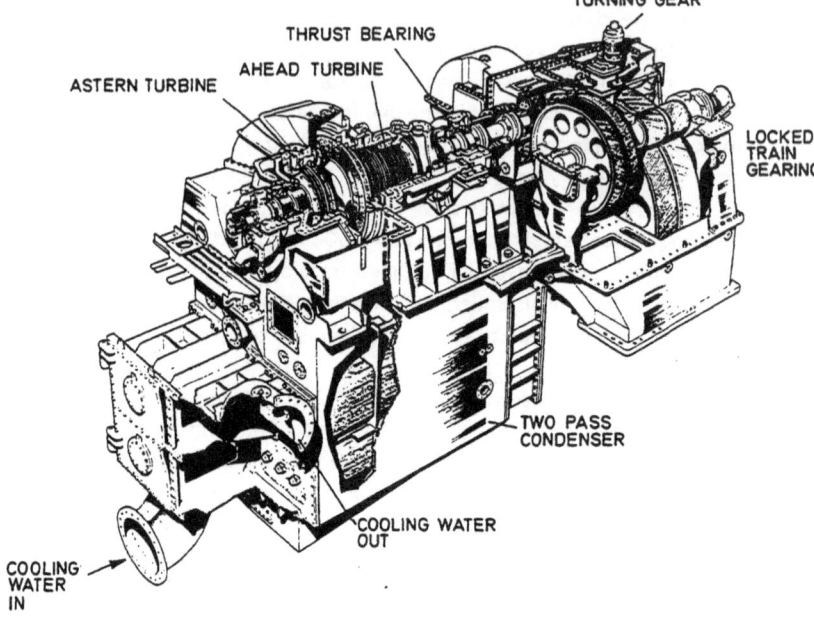

Fig. 4.11 CUTAWAY VIEW OF 20,000-SHP SINGLE-CYLINDER TURBINE PROPULSION MACHINERY

Fig. 4.12 shows a diagrammatic representation of the above turbine unit suitable for examination purposes.

Cross compound set

Basis plans for typical h.p. and l.p. units are shown in Figs. 4.13 and 4.15, with corresponding examination type sketches Figs. 4.14

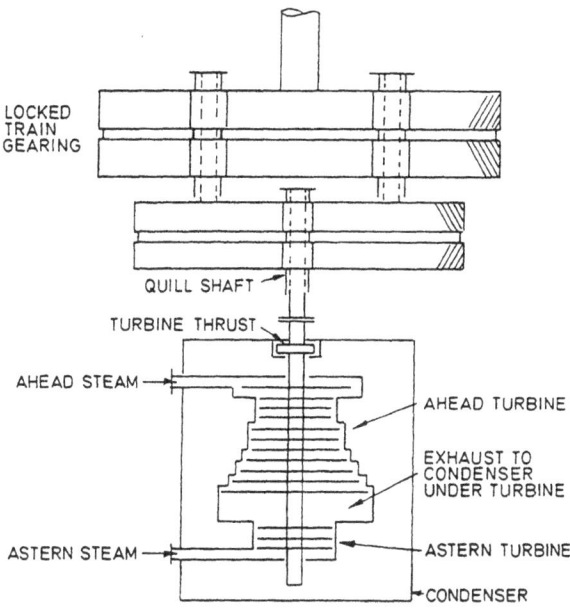

Fig. 4.12 SINGLE CYLINDER TURBINE & GEARING (15,000 KW UNIT)

and 4.16. The h.p. turbine, of double casing design, has an overhung astern turbine. The l.p. turbine is of the double flow type, double casing, actually mixed impulse and impulse-reaction. Similar arrangements for an all impulse-reaction design were very common but the trend is towards all impulse.

Alternative compound set

Referring to Fig. 4.17 it will be noted that the h.p. turbine is of simple casing design with diaphragm carriers. There is a two row Curtis wheel followed by nine single row wheels (Rateau stages) and the diaphragms are supported in plain grooves in the short inner carrier cylinder sections. Co-axiality of bearings, glands and diaphragms is by support palms, key and dowel pins. The design is claimed to be most suited to varying load and temperature conditions. The astern turbine is again overhung in this particular set. The l.p. turbine of Fig. 4.18 is a single flow design which is claimed to have a better performance

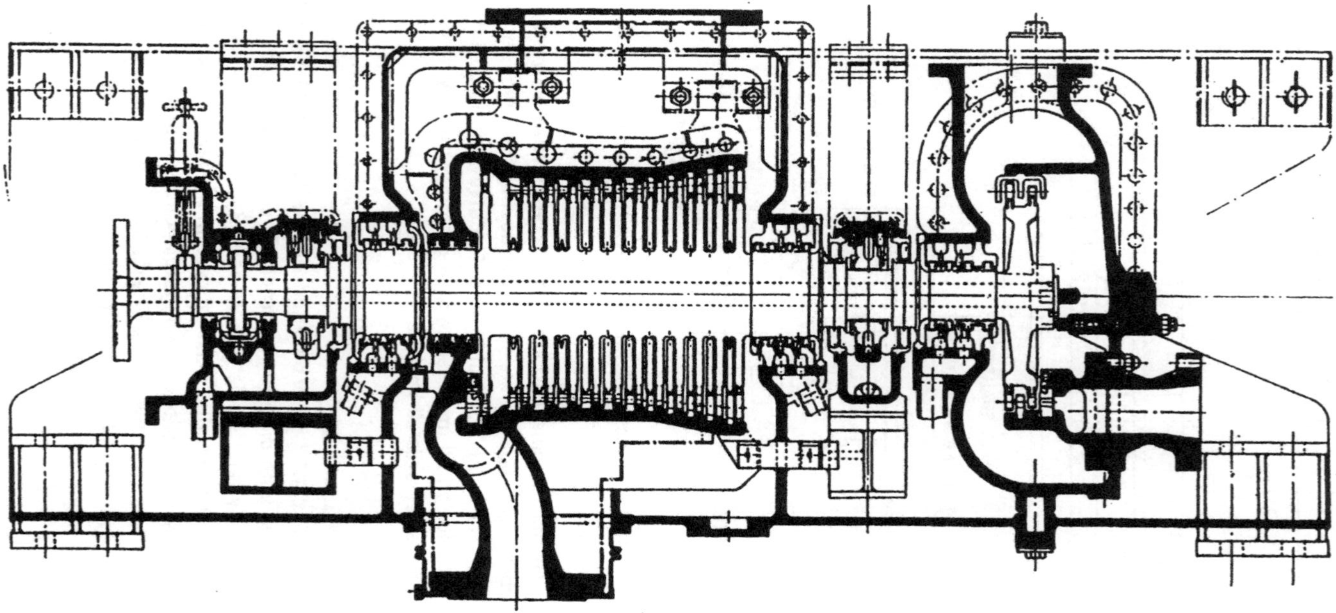

Fig. 4.13 H.P. TURBINE (IMPULSE) CROSS COMPOUND SET

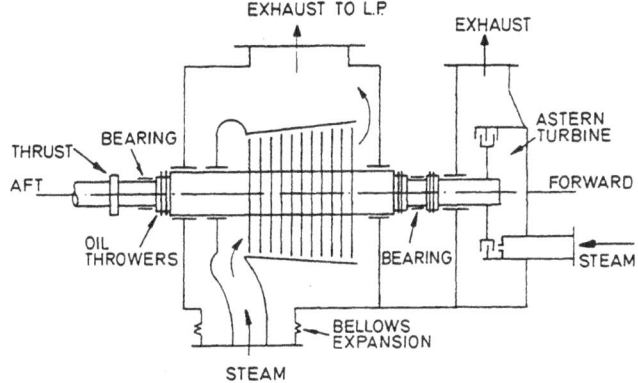

Fig. 4.14 H.P. DOUBLE CASING TURBINE

than the double flow type and due to closer support bearings be less liable to heat distortion. It is a double casing turbine with five impulse stages followed by four impulse-reaction stages (*i.e.* mixed, form decided by blade profile) for ahead and a two row impulse wheel followed by a single row impulse wheel for astern, in the same outer casing.

Reheat turbine

Fig. 4.19 shows in examination sketch form a reheat turbine. The following points should be noted:

1. The h.p. steam inlet at the centre of the turbine gives the smallest possible pressure and temperature differences across the casing and internal division (note the internal gland at the division).
2. The double flow will reduce the net axial thrust to be carried by the thrust bearing.
3. When warming through, the casing expands forward, whilst the rotor expands from the thrust aft. The effect of this is to reduce axial tip clearance in the h.p. turbine and increase it in the l.p. It should be remembered that clearances are more important at high pressures.

General construction

Referring to Fig. 4.20:

Rotors may be solid, hollow, built up or gashed. Gashed rotors are

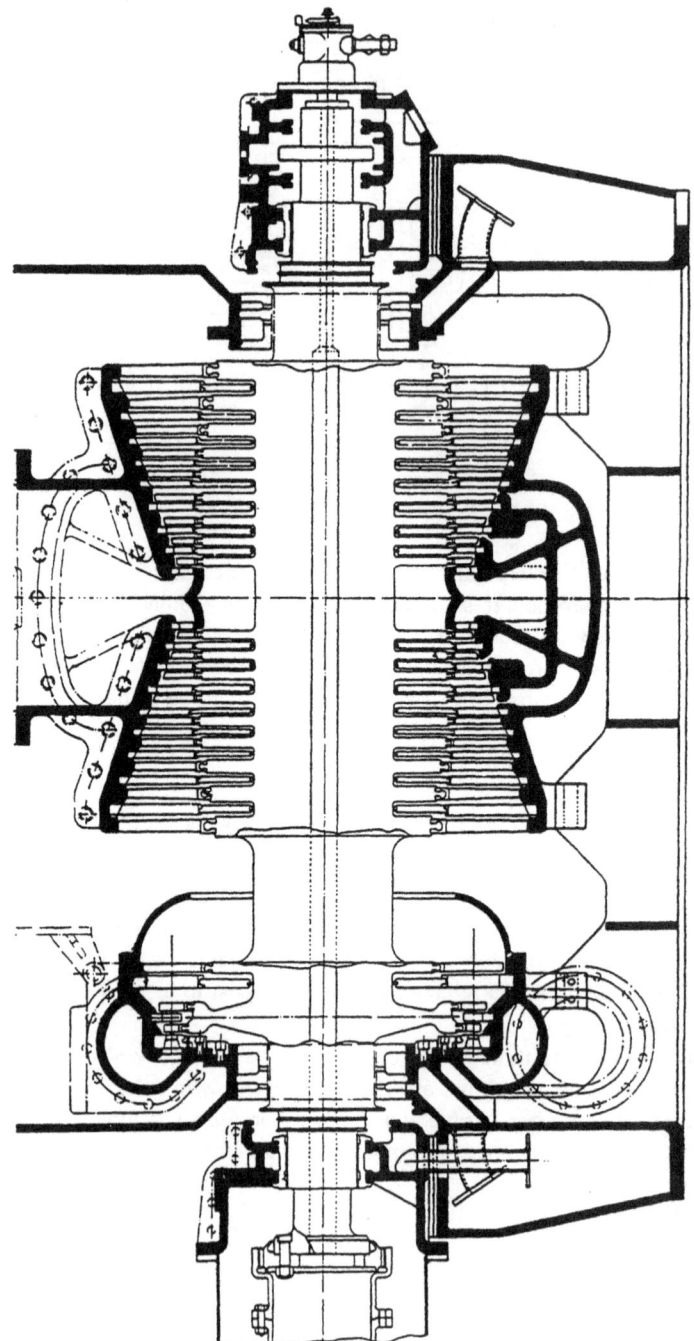

Fig. 4.15 L.P. TURBINE (IMPULSE) CROSS COMPOUND SET

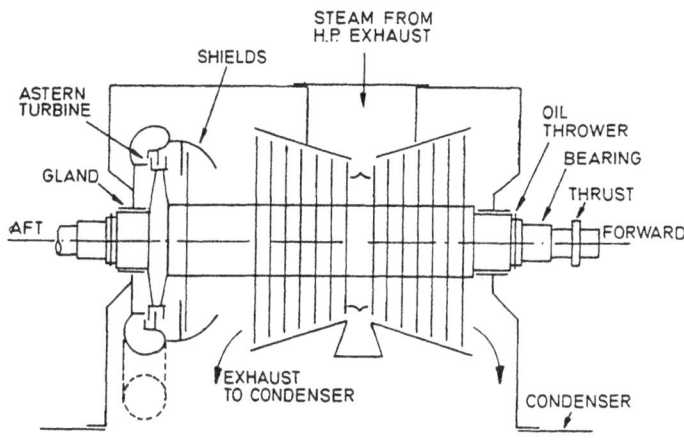

Fig. 4.16 DOUBLE FLOW L.P. IMPULSE TURBINE

mainly used in modern practice throughout but some l.p. rotors utilise built up construction, the sketch assumes a gashed rotor for the h.p. and built up for the l.p. as illustrated. For built up rotors the shaft is stepped (usually down from mid length on each side) and the wheels shrunk on and perhaps doubly or singly keyed. Gashed rotors (gashed out of the solid forging) allow the relative rates of expansion of rotor and casing to be more easily matched by correct proportioning and allow closer axial clearances which however does not affect efficiency greatly for an h.p. turbine.

The impulse turbine may have a velocity compounded stage followed by up to 12 pressure stages or as many as 16 pressure stages alone.

The wheels rotate in regions of almost constant pressure so axial clearance is not required to be reduced to a small amount. Inlet nozzles and fixed blades (for a velocity stage) particularly at the h.p. admission end only extend around an arc of the circumference of about 120 degrees, almost always in the top half of the casing, increasing in arc proceeding through the turbine to the l.p. Pressure equalising or balance holes, which are intended to eliminate end thrust, are rarely present in modern turbines as thrusts are small and the holes introduce stress and strength problems. The inspection hole is by no means standard. Blade heights, fixed blade annuli and nozzle

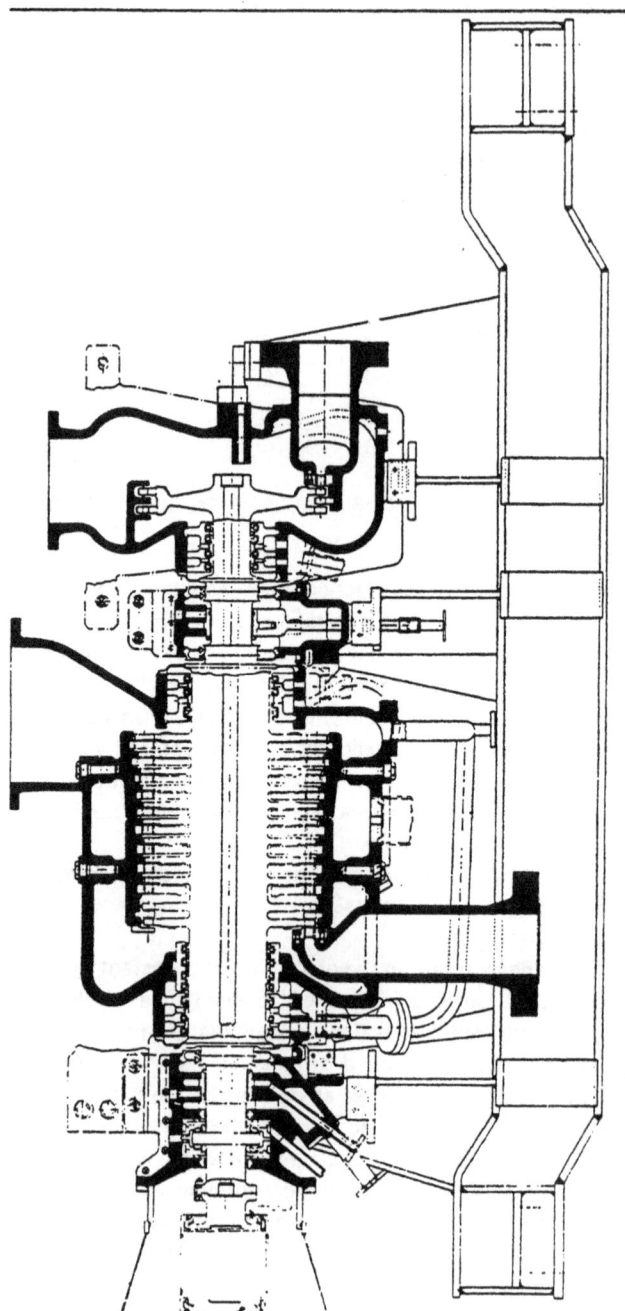

Fig. 4.17 H.P. TURBINE (IMPULSE) ALTERNATIVE CROSS COMPOUND SET

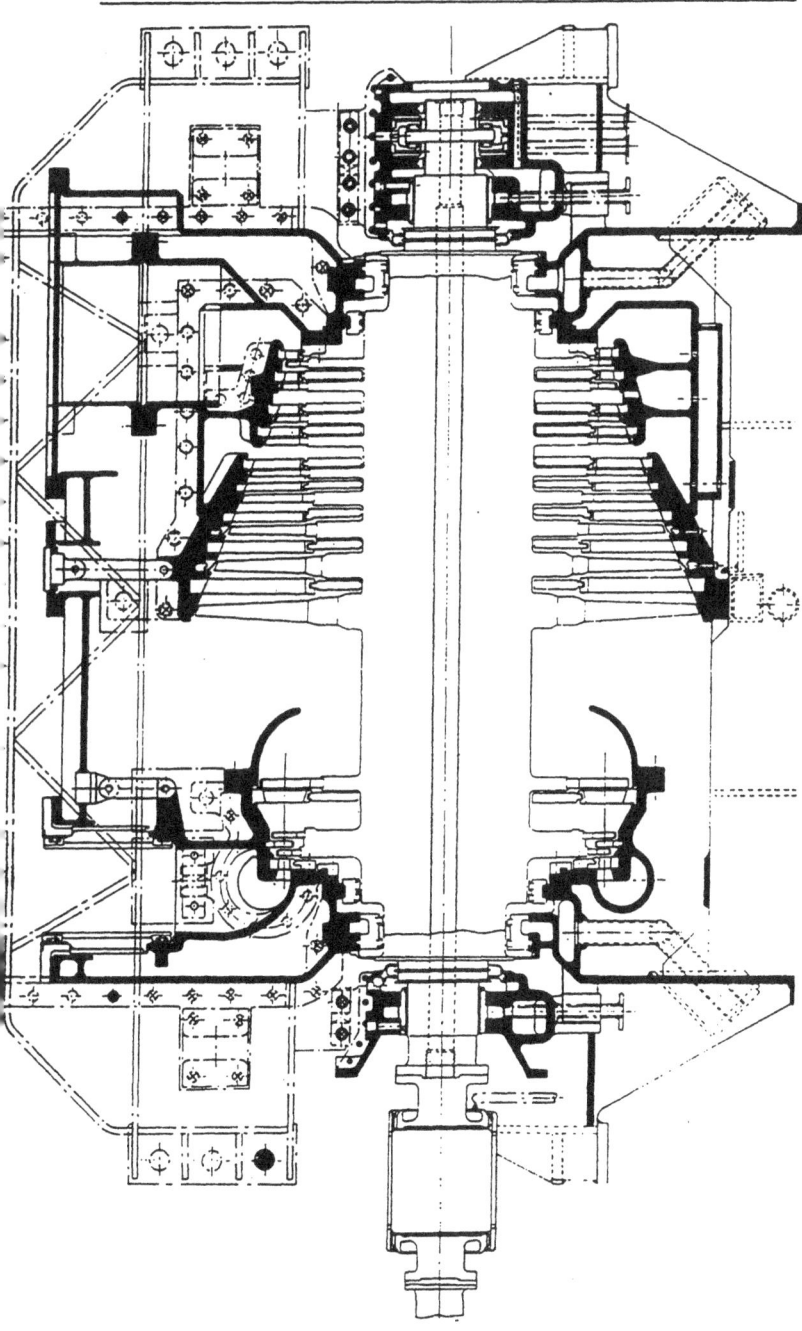

Fig. 4.18 L.P. TURBINE (IMPULSE) ALTERNATIVE CROSS COMPOUND SET

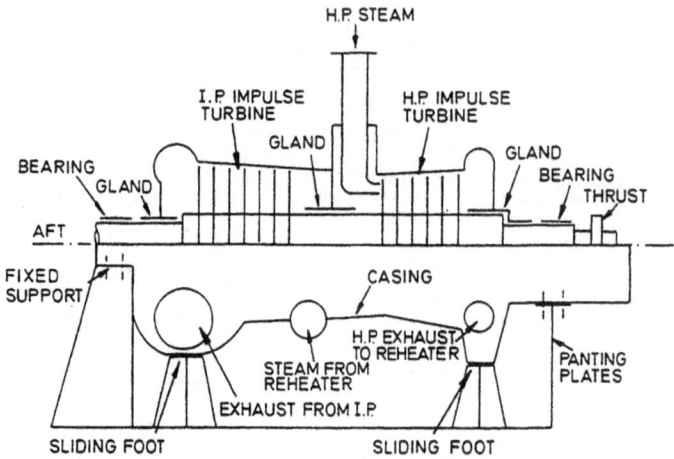

Fig. 4.19 REHEAT H.P.-I.P. TURBINE

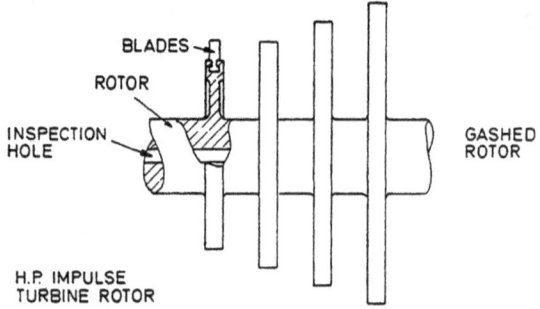

H.P. IMPULSE
TURBINE ROTOR

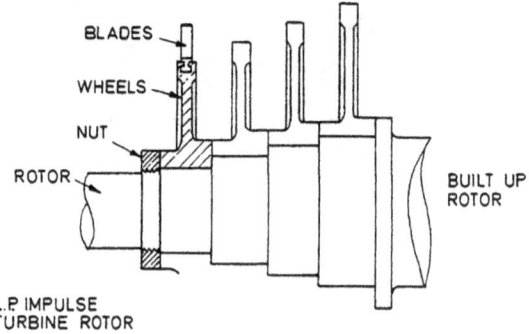

L.P. IMPULSE
TURBINE ROTOR

Fig. 4.20

areas increase progressively through the turbine. This is true even in a velocity stage, *i.e.* consider the following:

$$m = V \times A/v$$

m is mass flow, kg/s.
V is velocity, m/s.
A is area of flow, m^2.
v is specific volume, m^3/kg.

V is falling, v increases only slightly (as pressure is fairly constant), hence A must increase for constant mass flow.

Blading details

Definitions: 1. Segmental blading. Individual blades are assembled in a group together with distance pieces and shroud before they are fitted to the turbine rotor. Eight to twelve blades in a segment, the number being determined from vibration considerations. 2. Integral blading. Each blade is an entity, it has its own root piece and (in some cases) shroud. They are fitted singly.

Various types of blades and root fastenings are shown. Fig. 4.21 shows first the older type of blade form with the serrated root—a return is now being made to this type of construction, after some departure to straddle arrangements. However all types are possible.

A typical modern h.p. turbine arrangement would use 'T' root fastened blades inserted into a gateway, which is machined in the wheel, and packed around in the groove circumference of the wheel. Rolling the wheel flanks adjacent to the blades tightens them up. The closing blade fitted into the gate is riveted into position by a axial pin through root and rotor. Stainless steel shrouds, machined at inlet edge to give correct axial clearance, are fitted and the tenons projecting from the blade through the shroud are riveted or caulked over.

A typical modern l.p. turbine would have the first four or five stages of blading made up in the same way as the h.p. turbine described above. The last few stages of blades increase considerably in length because of the high specific volume of the low pressure steam. These latter blades invariably have 'variable geometry' *i.e.* they are tapered and twisted in section along their length. This takes into account changes in radial, tangential and axial velocities of the steam from root to tip.

Shrouding would not be fitted, even though it would be useful in

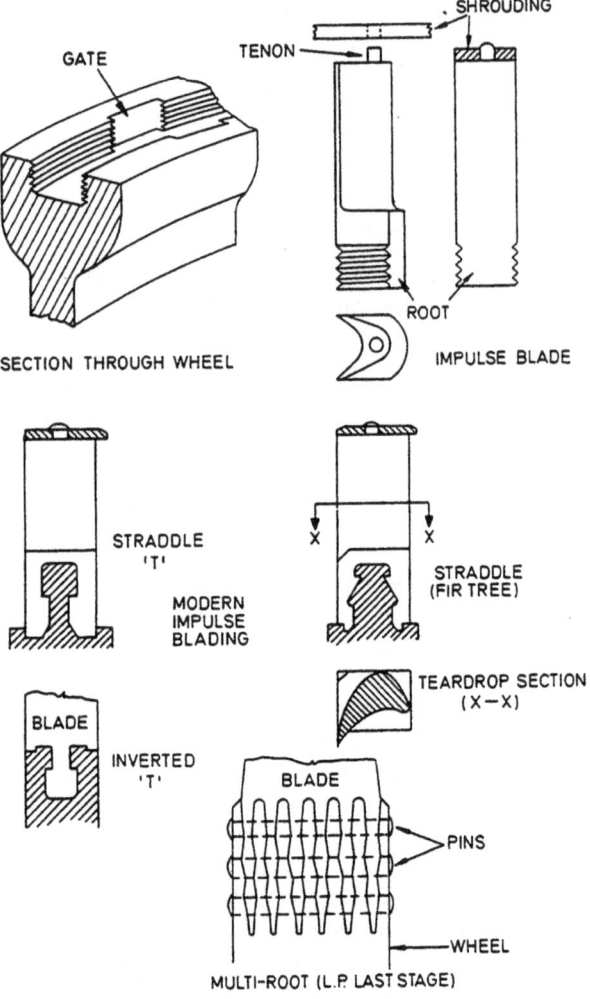

Fig. 4.21 IMPULSE BLADES AND ROOT FIXINGS

reducing tip leakage loss, because 1. the problems of fitting to awkward shaped blades, 2. the centrifugal stresses encountered would be very great. Instead the blades are knife edged with small radial 'tip clearance', if rubbing occurs the knife edge is worn down.

Root fixings for the last stages could be inverted fir tree and for the final stage pinned multi-finger fir tree root (Fig. 4.21). Basic reasons for departure from the simple inverted 'T', used in h.p. and first few stages of l.p. turbine, are: (a) very tight fit, does not depend upon centrifugal forces for tightness, (b) reduced possibility of fretting at low speed, (c) provides high strength under oscillatory bending action, (d) would not be used in h.p. stages due to increased machining and fitting cost.

Blade material should have: 1. Good creep and fatigue resistance, 2. High strength at high temperatures, 3. Good resistance to erosion and corrosion.

Some causes of turbine blade damage and failure
Vibration
This can cause fatigue failure. Blades are designed and fitted to keep vibration to a minimum. Causes of vibration:

1.   Running the turbine at other than the designed speed for prolonged periods may lead to blade failure if the blades are vibrating at resonance in a particular mode (a mode is a type of vibration).
2.   Steam excited vibration: partial admission could cause disc vibration. Nozzle passage frequency may lead to (1) above.
3.   Irregular steam flow pattern *e.g.* at a bled steam point.
4.   Hull transmitted vibration from other machinery.
5.   Gear misalignment or partial seizure of the flexible coupling may transmit periodic forces to the blades.
6.   Deposits of silica, sulphate, sodium or chlorides on the blades. These increase stresses and may alter blade frequency.

Lacing wires are fitted to some of the large l.p. turbine blades in order to prevent certain modes of vibration. Damping wires which pass through clearance holes in the blades and have their free ends peened over, minimise the amplitude of certain vibration modes by friction between the vibrating blade and wire. Type of root fixing can also alter fundamental frequencies of vibration as can the formation of segmental blading where the blades vibrate as a group.

Erosion
In order to obtain maximum efficiency it is necessary to expand the steam into the wet saturated region. Hence, in the last few stages of the l.p. turbine the blades are subjected to erosion by moisture. Due

to centrifugal effect the moisture is thrown radially out at the same time as it is moved axially with high velocity. Erosive effect is, therefore, greatest at blade tips and leading edges. To offset this, blades may have stellite or monel metal shields fitted to them by electron beam welding over two thirds of the leading edge from the tip. Or they may have flame hardened edges, note: 1000 m/s and 10% wetness would represent a rough limit for unshielded blades.

Water removal channels between adjacent diaphragms are also provided to prevent erosion damage. The water from the channels cascades, via axial holes of increasing diameter drilled in the bottom if the diaphragms, into the condenser or to a drain heater.

Operational troubles can lead to erosion problems *e.g.* superheat temperature too low, this results in wetter conditions in the l.p. turbine. Too high a vacuum or blocked drains.

Thermal distortion

There are several possible causes of thermal distortion of the turbine rotor. Incorrect warming through, incorrect use of gland steam causing gland rubbing, water entering the turbine resulting in 'thermal shock', steam leakage into astern turbine while running ahead etc. Thermal distortion could result in blade tip rubbing in the l.p. turbine which could strain the lacing wire brazing causing it to fail. This would alter the frequency modes of vibration and fatigue failure could result.

Nozzles and diaphragms

Nozzles are provided in groups to provide economic operation over the full speed range as throttling at manoeuvring valves is most inefficient, the nozzle plate is bolted to the casing and the formation of nozzles is much the same as that described below for diaphragms (for detailed discussion on this topic refer to control chapter). For high pressures and temperatures a bar lift type of nozzle valve operating gear is preferential where the correct nozzle setting at any power is obtained from one ahead and one astern handwheel.

In the case of diaphragms, expansion must be allowed for the diaphragm in the casing whilst allowing to expand and yet maintain the cold condition concentricities. Such an arrangement to allow radial expansion and maintain shaft gland clearances at a minimum is shown in Fig. 4.22.

The lower half of the diaphragm is supported in the lower casing by two lugs resting in the casing and recessed into the diaphragm. The upper half is supported by two plates slotting into the upper joint and recessed into the diaphragm. Centralising is done by the radial keys. Adjusting screws may be fitted in the diaphragm to bear on the lugs to allow vertical adjustment. The top key may be omitted if the top half of the diaphragm is keyed to the lower half to maintain the two in alignment.

Diaphragms tend to be of all welded construction. Nozzle forming blades are positioned by close fitting holes punched in thin steel bands forming the inner and outer walls of the nozzles to form the ring. These inner and outer bands are then butted up to the diaphragm centre inside the nozzle ring and the diaphragm periphery outside the nozzle ring, the whole assembly is then welded together to form a continuous structure. A modified non-welded design of diaphragm utilises the nozzles fixed into the casing by a strip and packer arrangement. The diaphragm centre rim is then simply supported at its periphery to the inner ends of the nozzle vane ring by a channel section shroud band riveted to the inner part of ring, this is shown in Fig. 4.22.

Single plane turbine plant

This plant depicted diagrammatically in Fig. 4.23 and shown sectionally in detail in Fig. 4.24 has the axes of the shafts, turbines and condenser arranged in one horizontal plane. This simplifies construction, reduces cost and gives about 35% reduction in foundation mass compared to conventional multi-plane arrangements.

The 8 stage impulse l.p. turbine exhausts axially into the condenser, this minimises exhaust loss (*i.e.* increases efficiency) and reduces engine height. Incorporated within the l.p. turbine casing is the astern turbine consisting of 2, two row velocity compounded stages exhausting axially in the same direction as the ahead turbine. This exhaust arrangement avoids heating of the ahead blading when running astern.

The all impulse h.p. turbine has a large diameter first stage, this gives relatively high peripheral speed and high operating efficiency irrespective of load. Following this are 8 impulse stages on a relatively small diameter rotor shaft—this reduces gland leakage.

Double reduction gearing incorporating epicyclic planetary first stage reductions on the h.p. and l.p. turbine shafts is used. The condenser

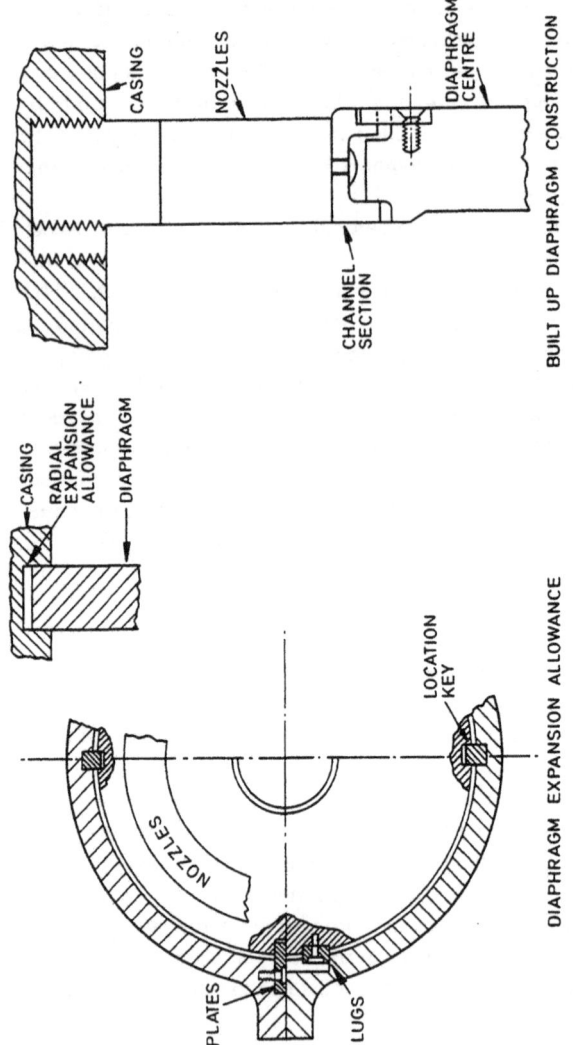

Fig. 4.22 IMPULSE TURBINE DIAPHRAGM DETAILS

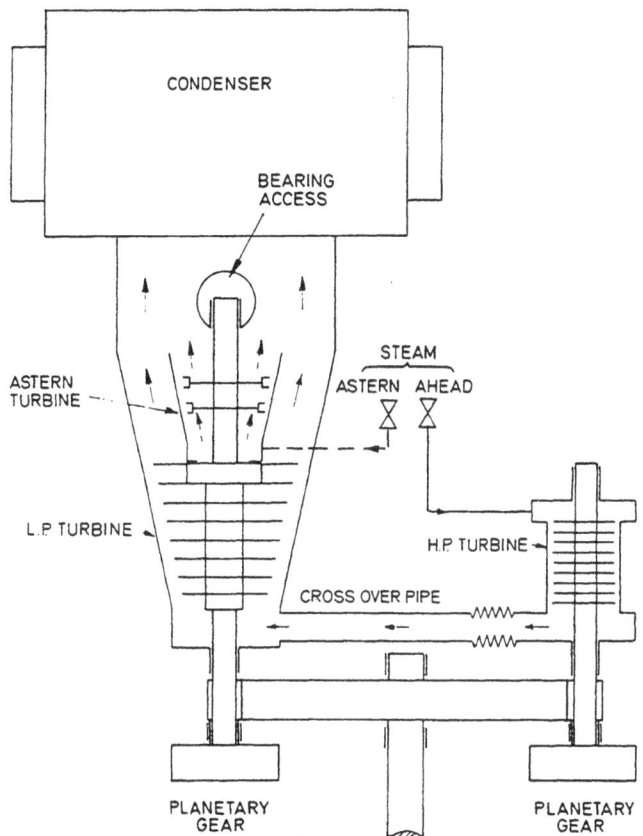

Fig. 4.23 SINGLE PLANE TURBINE PLANT

may be double or single pass, the latter being used with scoop intake—this minimises resistance.

IMPULSE-REACTION TURBINES

Such turbines are generally called reaction turbines but both impulse and reaction principles are utilised in the blade profile. Impulse-reaction turbines were almost standard practice up to 1945.

The increase of pressure and temperature in modern practice has

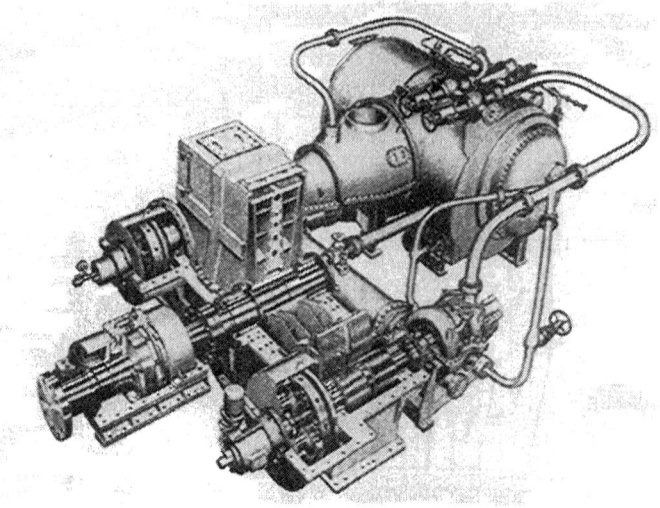

Fig. 4.24

meant a movement towards impulse turbines. The efficiency of impulse-reaction turbines is critical to minimum blade tip leakage and this becomes difficult to arrange at elevated pressures. Hence the descriptive work on impulse-reaction turbines has been somewhat reduced to main essentials in this section. For low pressures the impulse-reaction design is still often used, particularly for l.p. turbines.

In the three cylinder set, h.p., m.p., l.p., the h.p. astern turbine (usually a simple impulse wheel) is mounted on the end of the m.p. ahead unit and the l.p. ahead and astern turbines are single flow, on the same shaft. Typical pressures and temperatures utilised for such a set would be: h.p. inlet 18 bar, 288° C, m.p. inlet 5 bar, 175° C, l.p. inlet 0.4 bar, 88° C, h.p. and m.p. turbines will first be described with details and then the description and details of l.p. turbines considered.

H.p. and m.p. turbines

A typical design is shown in Fig. 4.25 with a basis plan. In the detail sketch the rotor is shown of hollow construction but a solid rotor can be fitted. A two row Curtis wheel is fitted before entry to the h.p. turbine, this is often fitted at entry to all h.p. impulse-reaction turbines and in some cases to the entry to the m.p. turbine, such a wheel is also provided for the astern turbine or turbines. The main object of such a wheel at h.p. entry is to drop the pressure by doing work

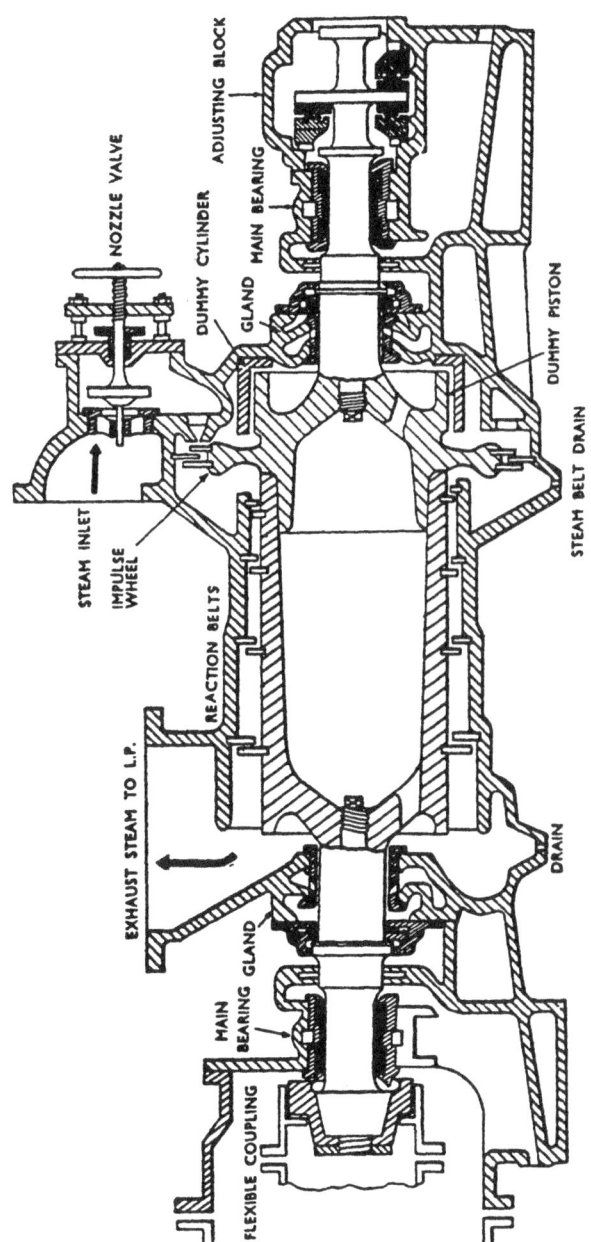

Fig. 4.25 H.P. TURBINE (DETAIL). IMP/REACTION

through the impulse wheel so providing inlet steam to the start of the reaction belts at lower pressure which means less blade tip steam leakage will occur through the reaction stages, also gland sealing is less difficult. Nozzle control valves are shown to give speed control but many impulse-reaction turbines just have the nozzle entry from the manoeuvring valves directly without control valves.

Ideally, with the impulse reaction design, the blades should progressively increase in height from blade row to succeeding blade row as steam flows in the axial direction. In practice to avoid a complex casing curve form and numerous blade heights the casing is stepped in stages with a set of blades of the same height in each reaction stage. About 12 blades in each of about five reaction stages is often utilised for one turbine. Note the standard form of flexible coupling, bearing, glands, drains, etc. The dummy piston, equalising pipe, blading, and adjustable thrust block details together with the axial clearances utilised are deserving of particular attention and will now be discussed separately.

Dummy piston

An extension of the rotor cylinder is provided and this runs inside the dummy cylinder (see Fig. 4.25). Packing rings are grooved into cylinder and piston. This acts as a double gland and so reduces pressure on the spindle gland. Such a piston is normally only fitted at entry to h.p. turbines. In some cases a similar dummy may be fitted at the exhaust end but this is not common. The gland packing used for the inlet dummy as sketched is of the facial type as described previously under external turbine glands (see Fig. 4.10).

Equalising (balance) pipe

This is shown in Fig. 4.26. The object of this pipe, or pipes, is to steam balance the rotor axially so tending to relieve the rotor axial steam thrust caused by steam force on the blades. If pressures are equalised at forward and after ends by the connecting pipe and the areas are arranged to be approximately equal, then steam forces are balanced. The dummy piston thus serves an equally important function of providing the area under balanced pressure at the forward end, hence the term piston. On Fig. 4.26 the balance forces of pA are where illustrated. Not all turbines have such balance pipes as the area and steam pressure can be arranged to give approximate balance without a connecting pipe.

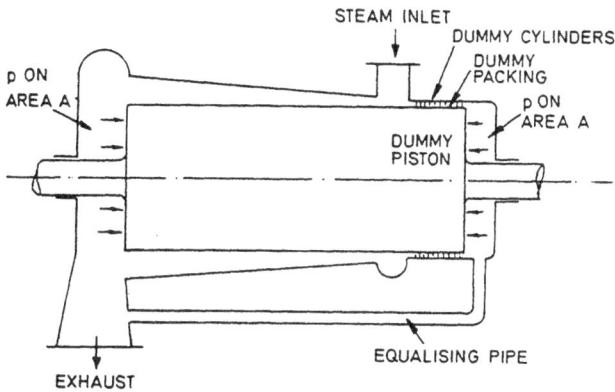

Fig. 4.26 DUMMY PISTON & EQUALISING PIPE

Blading details

The construction of impulse-reaction turbines consists of alternate rows of fixed and moving blades and the layout arrangement should be clear from Fig. 4.25. As pressures in h.p. and m.p. turbines are fairly high, clearances at blade tips must be reduced to a minimum for good efficiency. This is best achieved with an *end tightened* form of construction. Clearance is in an axial direction which is easier to arrange than in a radial direction. The axial clearance is also adjustable so that clearances can be increased for manoeuvring when temperature fluctuations could cause expansion variations with metallic contact, whereas uniform full load operation under steady temperature conditions can utilise minimum clearances for maximum efficiency without any dangers of metallic touch. Nonetheless the axial position of the rotor becomes most vital and regular checks during overhauls, when warming through and when running are essential (clearance gauges have been previously detailed, see Fig. 4.3).

Referring to the details given in Fig. 4.27:

The side packing pieces are of soft iron and the whole serration and edges are fully caulked in place after fitting. The approximate running clearance for h.p. and m.p. turbines is about 0.33 mm when running (minimum) which is increased to about 1.1 mm for manoeuvring (see later). Shrouding is ground to a knife edge at possible contact point. Segmental blade fitting is now almost always adopted. About

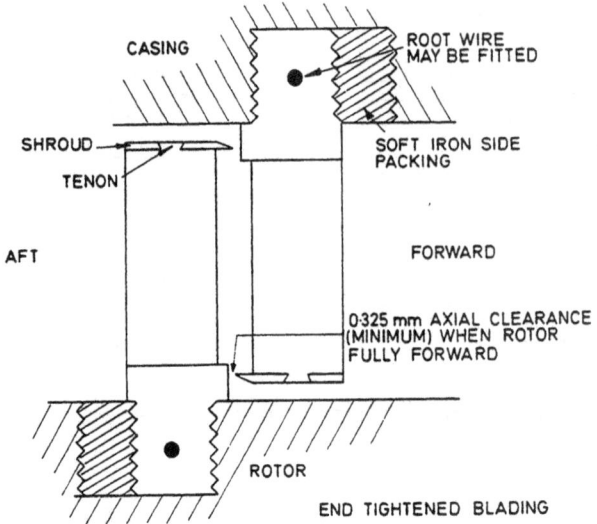

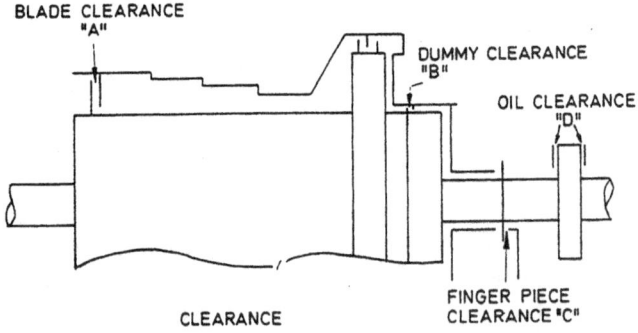

Fig. 4.27 CLEARANCES

10 blades are in a segment section with the last few blades at each end secured by a root wire and end caulked whilst in a mandrel. Modern practice utilises integral blade and root in place of separate components as in the top sketch and utilises molten brazing or electric welding in the mandrel. The segment is finally caulked in place with the end tightening strip alongside in the rotor or casing groove,

this is completed for all segments around the periphery. Tip tightened blading is utilised in l.p. turbine practice and will be described later.

Clearance details

A brass plate is usually fitted on the casing with a permanent record of finger piece readings as taken during manufacture completion gauging. The finger piece is fitted to the casing near the gland by countersunk screws and feeler gauges inserted between finger piece and a collar on the shaft. These readings are always taken for comparison before lifting the casing and after lowering as a check and to ensure no fouling risk may occur. The condition of turbine, *i.e.* hot, warm, cold, etc., should be specified. A typical set of plate details are given below for an h.p. turbine:

	Rotor Forward	Rotor Aft
Thrust Cage Forward	0.25 mm	0.65 mm
Thrust Cage Aft	1 mm	1.4 mm

A screw and nut arrangement on the rotor end allows it to be moved axially. With the thrust housing fixed either forward or aft the rotor travel is 0.375 mm, this must be the total oil clearance between pads and thrust collar, *i.e.* 'D' in Fig. 4.27. The travel of the thrust cage is 0.75 mm which is the amount the rotor can be moved aft to increase blade clearances for manoeuvring.

Now if it is recorded, that with the pads out and the rotor drawn forward so that the blade shrouding is touching, the finger piece reading is 0.125 mm and the dummy packings are (on average) about 0.125 mm clear then the position of the rotor when running can always be established. Again referring to Fig. 4.27 and considering the finger piece reading when running is 0.45 mm (see C). Dummy clearance (see B) is then 0.45 mm, blade shroud clearance (see A) is 0.325 mm. Oil clearance forward is 0.2 mm and aft 0.175 mm. The position of rotor is thus fully established, all clearances are also known.

The above is typical only for a given turbine, m.p. turbine readings would be fairly similar.

Adjustable thrust block

The thrust must be movable axially to increase axial clearances for

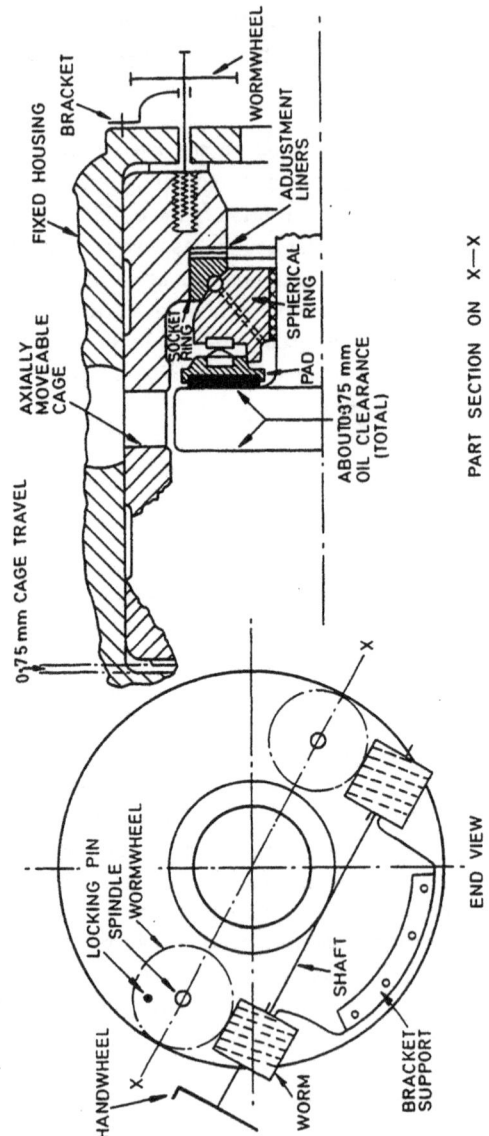

Fig. 4.28 THRUST ADJUSTMENT WITH END TIGHTENED BLADING FOR IMPULSE-REACTION TURBINES

manoeuvring by about 0.75 mm, *i.e.* in example above blade clearances 0.325 mm to 1.075 mm. This is arranged by having the thrust casing made as a cage which can be moved axially inside the outside fixed housing (see Fig. 4.28).

A conventional thrust block can be moved by a screw arrangement (at two points) as shown. From 'Contacts On', *i.e.* rotor forward, remove the locking pin and rotate the handwheel which rotates the worm and wormwheels. These fittings being external and forward of the fixed housing and bolted to it by a bracket. The rotation of the wormwheel spindles moves the cage back 0.75 mm, the locking pin is now replaced in the new position 'Contacts Off'.

This operation is done before manoeuvring with the turbine running and the description should be carefully considered alongside the sketch of Fig. 4.28.

L.p. turbines

A double flow design is sketched in Fig. 4.29 as a detail basis plan. Hollow rotor construction is most common, single flow designs of three cylinder turbine sets almost identical in layout plan—except for the blades and double casing—as the l.p. impulse turbine (alternative cross compound set) given in the basis plan of Fig. 4.18 were almost standard practice. These tended to be replaced by two cylinder sets of double flow design in the l.p. of which Fig. 4.29 is typical except that nozzle control valves for astern turbines are not a common arrangement.

Impulse-reaction turbines are much more suited to low pressures as blade tip leakage losses are much reduced. However, impulse turbines of double casing construction are now being used in the single flow form more and more for l.p. practice to replace such reaction turbines. The double flow design is often criticised as being subject to too much heat distortion caused by the rather lengthy bearing distances. American practice utilises astern wheels at each side to give symmetry and more even temperature distribution. Length, however, is slightly increased. The single flow design is claimed to give a higher thermal efficiency.

Referring to Fig. 4.29:

The only aspects of construction which differ from h.p. or m.p. practice or general turbine details previously described are: (a) There

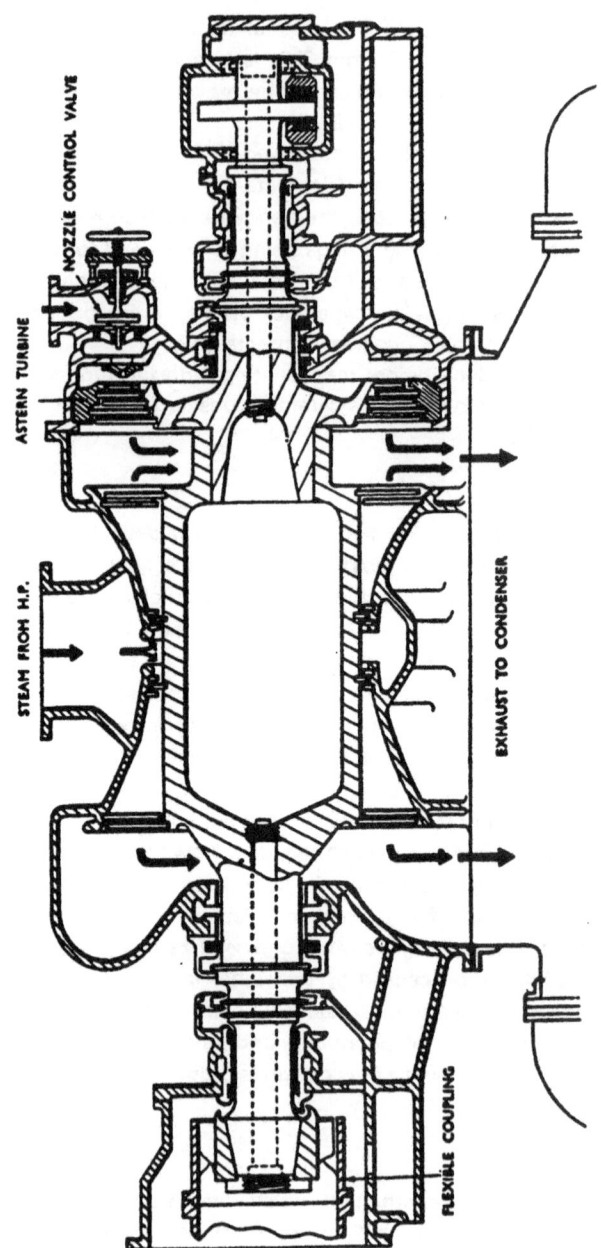

Fig. 4.29 DOUBLE FLOW L.P. TURBINE (DETAIL). IMP/REACTION

NOZZLE CONTROL VALVE

ASTERN TURBINE

STEAM FROM H.P.

EXHAUST TO CONDENSER

FLEXIBLE COUPLING

is no adjustable thrust block and (b) End tightened blading is not utilised.

As previously mentioned, at lower pressures the differential pressure across the blades is not high and steam leakage is low so that the more expensive end tightened blades with attendant thrust adjustment block are not utilised. Tip tightened blading, with radial clearances, is normally used.

Blades may be integral or with separate blade and root although the first is more common. The blades are caulked directly in place after segmental forming. Serrations and side packers may also be used, likewise a root wire may be fitted. Knife edge radial clearance varies from about 0.5 mm to 1.75 mm. Shrouding may be fitted with a right angled turn so that the shroud provides the knife edge clearance. Alternatively a binding wire of brass or copper was used in early designs. More modern practice utilises alloy wire often silver soldered to the blades, for high blades occurring in the last stages of l.p. turbines the binding wire is often duplicated to give support to tip and mid length. Vibration is a difficult problem on long blades, this can be at a steam excited frequency.

SOME DESIGN AND OPERATION CONSIDERATIONS

Turbine choice is governed by some of the following points:

1. Power requirement for the desired ships speed. Power $\propto$ (ship's speed)^3 and present day trends are for higher speeds and quick turn arounds.

2. Steam pressures and temperatures. As steam pressure increases specific volume (*i.e.* m^3/kg) decreases, therefore nozzles and blades become smaller and less efficient. A pressure limit is reached for every power capacity of turbine when the gain due to higher pressure is offset by the decrease in internal efficiency. Higher pressures may be used more effectively in highly rated turbines, Fig. 4.30.

Vacuum is roughly standard at 0.05 bar (barometric pressure 1.013 bar), it will vary being slightly reduced with reduced sea water temperature.

The main governing factor is the specific volume of the steam, Fig. 4.31 shows the variation with pressure. In order to accommodate increase in volume the flow areas at exhaust must be proportionally increased, hence an obvious practical limit is arrived at.

Increasing initial steam temperature without altering the pressure will reduce steam flow rate (kg/h) through the turbine for a given power since heat content (kJ/kg) will have increased. Machinery dimensions are only altered a small amount to accommodate the slightly higher specific volume of steam resulting from increased temperature. For any initial pressure there is a minimum temperature below which the moisture content of the steam at the low pressure region of the turbine causes excessive blade erosion and efficiency loss. Often, dryness fraction of 0.88 at exhaust is the accepted norm (*i.e.* 12% moisture content).

3. Steam cycle, the number of bleed points and steam flow from them, reheating and feed heating, etc. all play an important part, this is examined in greater detail in the boiler and feed systems chapters.

4. Transmission system to the propeller, the steam turbine is essentially a high speed machine and the propeller is more efficient at

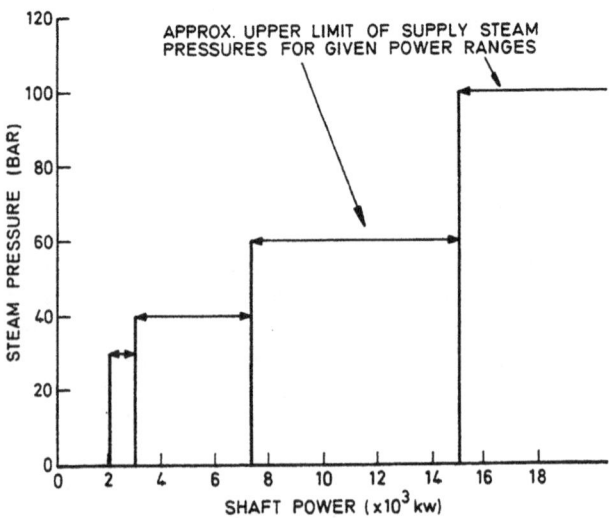

Fig. 4.30

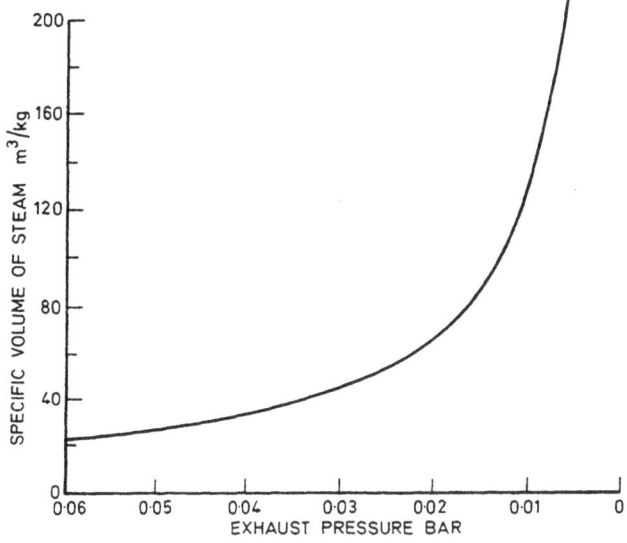

Fig. 4.31

low speeds. The greater the speed reduction the less efficient the transmission system of gearing, obviously triple reduction is more expensive and less efficient than double reduction, but the gain in propulsive efficiency may possibly offset this.

5. Astern operation, the astern turbine is generally integral in some way with the low-pressure turbine. It may be running in its own casing or it may be incorporated in the same casing as the ahead turbine, the latter arrangement reduces efficiency by about 1 to 2%, but has the following advantages: Decreased space and weight, lower initial cost, simpler piping and seating.

6. Space limitations and weight. Single plane systems, where the steam exhausts axially into the condenser. Gearing and turbine axes lie in the same plane, give reduced height but increased dimensions in the other directions. Gearing and condenser are the largest items to consider, use of locked train gearing and under-slung condenser (or l.p. turbine supported on the condenser) lead to a compact but relatively higher plant.

7. Fuel economy—this is likely to become of paramount importance. Methods of reducing fuel consumption are:

(a)   Use of cross compound turbine plant in favour of single cylinder.

(b)   More complex steam cycles with up to five stages of regenerative feed heating and reheating.

(c)   Scoop intake of condenser cooling water.

(d)   Improved and more reliable automation equipment—leading to on-line computer control.

Fig. 4.32 shows in a simplified form the amount of energy available for propulsion purposes—it is a relatively small amount compared to that going to the condenser, it is here where designers will possibly turn their attention in order to recuperate in some way the enormous amount of energy at present going to the condenser cooling water.

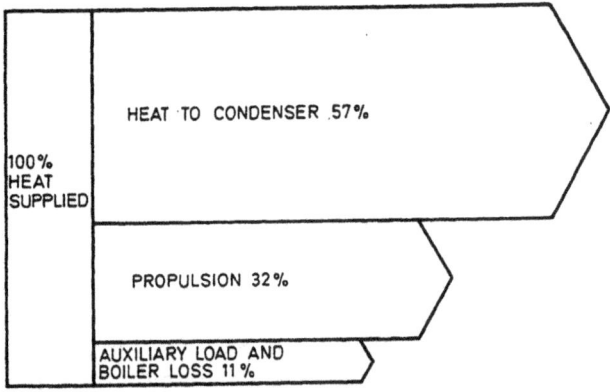

HEAT TO CONDENSER 57%

100% HEAT SUPPLIED

PROPULSION 32%

AUXILIARY LOAD AND BOILER LOSS 11%

Fig. 4.32 ENERGY FLOW DIAGRAM

Balancing of turbine rotors

To obtain vibration free running of a turbine rotor it should be: 1. statically and dynamically balanced, 2. operated at a speed in excess of its first fundamental critical speed (ω_n), but below its second critical speed $(4\omega_n)$.

Static balance is achieved when the centre of mass of the rotor is on the axis of rotation hence when the rotor is at speed there will be no out of balance rotating force vector. A rotor can be statically balanced yet unbalanced dynamically due to couples formed by equal opposing forces in different planes. Fig. 4.33 shows two discs with masses m_1 and m_2 at radii x_1 and x_2 respectively. With the rotor at rest the turning moments due to gravitational attraction will be balanced if $m_1x_1 = m_2x_2$. At speed the forces F due to these masses will balance but since these forces are in different planes, a rotating out of balance couple Fz will cause vibration.

The first fundamental critical (or whirling) speed ω_n (ω rad/s, suffix n for natural, since the circular frequency of natural transverse vibrations corresponds to the whirling speed) for a balanced rotor depends upon flexibility of rotor and bearings. The greater the stiffness of either, the greater the value of ω_n.

With modern turbines it is not practical to have ω_n greater than the normal running speed, hence during run up to operational speed the critical must be run through—this as rapidly as possible.

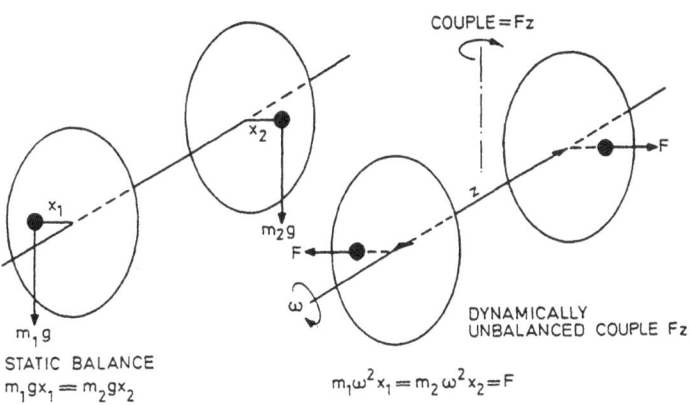

COUPLE $= Fz$

x_2

m_2g

F

z

ω

DYNAMICALLY
UNBALANCED COUPLE Fz

x_1

F

m_1g

STATIC BALANCE
$m_1gx_1 = m_2gx_2$

$m_1\omega^2 x_1 = m_2\omega^2 x_2 = F$

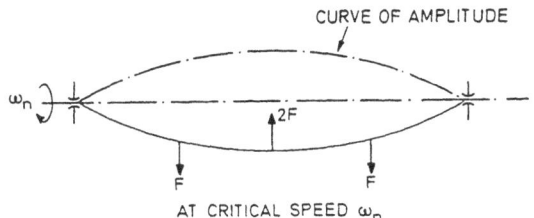

CURVE OF AMPLITUDE

ω_n

$2F$

F   F

AT CRITICAL SPEED ω_n

Fig. 4.33

The disadvantage of having to run through the whirling speed is the possibility of excessive vibration with damage. To obviate this, masses can be added to the rotor in such a way that whilst its balance is in no way affected the amplitude of the vibration at critical speed is reduced Fig. 4.33.

The advantage of operating at speeds greater than the first critical but less than the second is that the centre of mass of the rotating system moves towards the axis of rotation (*i.e.* no deflection due to weight) and this gives extreme stability (like a spin dryer or centrifuge).

ASSOCIATED EQUIPMENT DETAILS

Gland sealing steam system

Ideally the gland sealing system should ensure, at all times, no steam leakage from the turbine to the atmosphere, this would waste water and energy, and prevent air entry into the turbine. Air must be removed mechanically and this again means energy expenditure.

Fig. 4.34 shows a modern gland sealing steam arrangement, the high pressure end of the h.p. turbine having three inner gland pockets and one sealing pocket, whereas all other glands have one inner pocket and sealing pocket.

When a vacuum has to be raised sealing steam to the glands is supplied from the controlled valve A at 1.35 bar, the valve would be operated by remote control the auto control being by-passed, and distributed to the inner gland pockets.

Under running up conditions steam leaks from the inner pockets of the h.p. turbine and the after l.p. to the forward inner pocket of the l.p. where a partial vacuum exists, any make up steam required is supplied through valve A which would be on automatic, being controlled by the steam pressure in the turbine crossover pipe.

At full load there is an excess of steam from the inner pockets and this is bled off through the overflow valve, which automatically opens at a pressure of about 2.2 bar (this pressure can be regulated), this excess steam then passes to the gland steam condenser wherein air is extracted, and the latent heat is given to the boiler feed—this being the cooling medium.

Steam leakage from the inner pockets to the outer sealing pockets

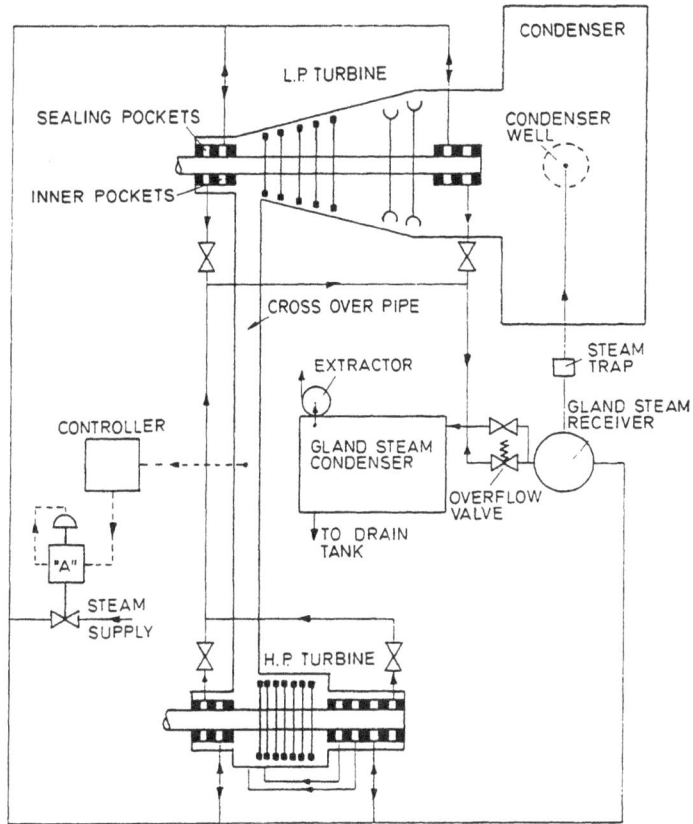

Fig. 4.34 GLAND STEAM ARRANGEMENT

also passes to the gland steam condenser, control of the steam pressure in the sealing pockets is achieved by valves. When adjusting the pressure the valves are first closed, this causes steam to leak out of the turbine, and then they are gradually opened until the leakage ceases.

Emergency steaming connexions

For normal operation

Emergency pipe the l.p. turbine would not be connected, blanks A and D would not be fitted. Blanks B and C would be in place.

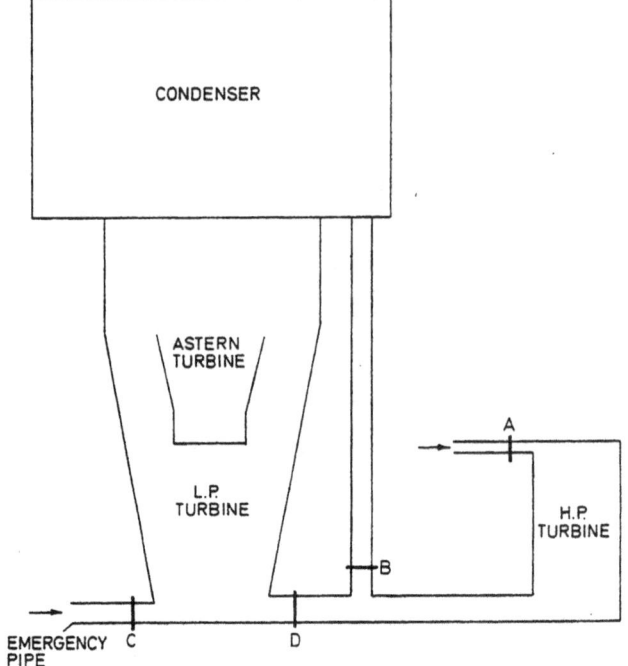

Fig. 4.35 EMERGENCY STEAMING CONNECTIONS

h.p. turbine inoperative

Disconnect h.p. turbine high speed coupling to the gearing. Fit blanks A and D, remove blank C and connect emergency steam pipe to the l.p. turbine. De-superheated steam can now be supplied to the l.p. turbine, and normal steaming can be used for the astern turbine.

l.p. turbine inoperative

Disconnect l.p. turbine speed coupling, fit blank D, remove blank B and lock astern steam valve in position. Operate at reduced steaming rate through the h.p. turbine, remember, no astern power is available as the astern turbine is integral with the l.p.

For a three cylinder cross compounded set the following alterations to the steam flow can easily be visualised.

1. Cutting out the h.p. ahead turbine by blanks and leading the

steam to the m.p. turbine using one portable pipe.

2.   Cutting out the l.p. turbine by blanks and leading the steam direct to the condenser (this can be done to l.p. ahead or l.p. astern or both).

3.   Combining 1 and 2 above to utilise m.p. turbine only.

4.   Using the l.p. turbine only by blanking and leading the steam to the l.p. directly using the other portable pipelength.

Conditions of reduced pressure and temperature adopted will depend on the arrangement used.

The drains are manually operated by valves with extended spindles. In general all drains would be shut at full power and gradually opened as speed is reduced, starting with l.p. drains, so at stop all drains would be full open.

Turbine Lubrication Systems

These may be either Gravity or Pressure, modern trend is towards the latter since: no large gravity tanks are required, piping is reduced, oil quantity is minimised—which all leads to a considerable economy. However, a safeguard must be incorporated to ensure no bearing run out in the event of oil pump failure. With the gravity system it is simple—oil flows from the gravity head storage tanks until the turbines come safely to rest. With the pressure system an auxiliary standby pump could cut in automatically but that would not be satisfactory in the case of total electrical blackout—so a main engine driven pump, and small gravity tank at a low level, may be a better and safer arrangement.

Gravity system

A typical system is as shown in Fig. 4.36.

The oil is drawn from the sump via a suction filter and magnetic filter by the pump discharging via a discharge filter through the cooler to pressure distribution piping, pressure controlled by a screw lift valve. A certain amount of oil passes through the valve to maintain the gravity tanks constantly overflowing through a sight return. The system is connected to storage tanks and purifiers, a connection leads to turbine protection devices. In the event of total pump failure oil can pass unrestricted through the non-return valve to distribution points, this usually gives emergency supply for about three minutes at

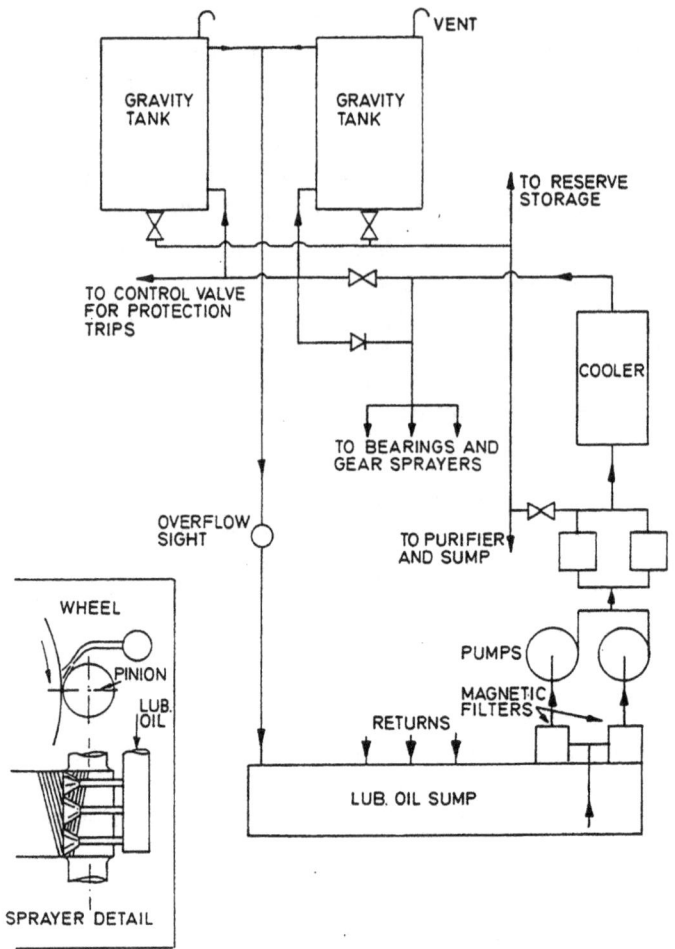

Fig. 4.36 GRAVITY SYSTEM

1.7 bar (normal supply pressure 3.1 bar), emergency shut off steam valves normally function at below 2 m oil head at the control valve unit. Pump failure alarm and standby pump cut in switches are normally fitted in addition to low level alarm devices on the gravity tanks. A detail of oil sprayers is given, oil supply is on inlet side of gear mesh but modern practice now tends to supply at both sides. The best oil inlet temperature is about 46.5° C.

Pressure system

Oil is drawn from the sump by the main independent and shaft driven pumps, which share the oil supply to bearings and gears.

In the event of main pump failure—due to electrical blackout for example, the shaft driven pump will continue to deliver. When low revolutions are reached the oil pressure falls and the small gravity tank incorporated in the top of the gear case takes over the supply to the system. Total run out time is about 25 to 30 minutes, which gives operators sufficient time to start the standby pump or stop the machinery.

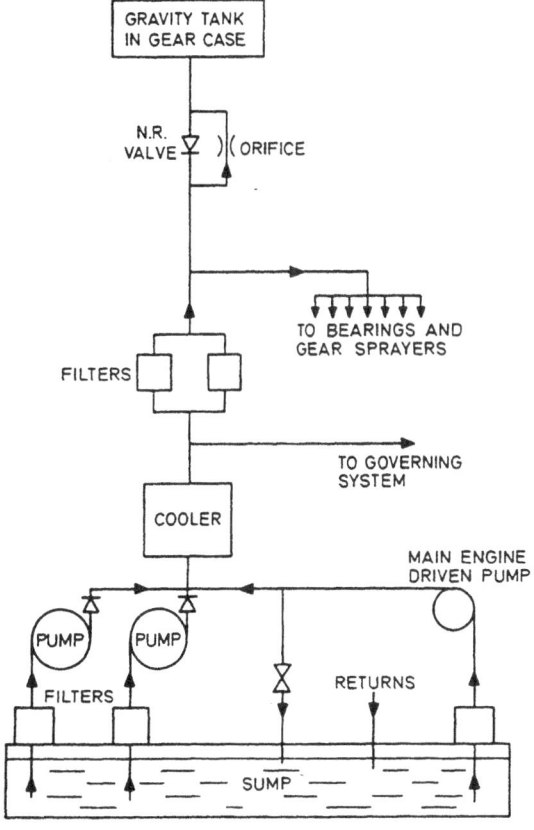

Fig. 4.37 PRESSURE SYSTEM

Care and testing of lubricating oil

The following points are important to ensure the lubricating oil is kept in good condition:

1.   Leakage of water into oil from gland steam and water pipe glands, etc. must be prevented.
2.   Maintain correct oil temperature to avoid condensation of water vapour.
3.   Maintain correct oil pressure, too low a pressure could mean insufficient oil passing through bearings, etc. with overheating of the oil resulting.
4.   Check regularly temperature differences across the oil cooler to ensure correct heat transfer.
5.   Check bearing temperatures regularly.
6.   Ensure ventilation system is in good order.
7.   Operate oil purification system on continuous by-pass.
8.   Watch condition of filters, examine entrapped solids.
9.   Take regular samples of oil, they should be drawn from an oil return line and be clear, free from water and sludge.

Shipboard tests could be carried out to estimate:

(a)   Water and salt content.
(b)   Viscosity.
(c)   Total and strong acid numbers.

The drop test is one which is frequently used and widely understood to give simple, reliable results. Indications of departure from new unused oil being the yardstick.

Oil coolers may have sludge and other deposits which have to be removed. Steam, hot water or a bath of trisodium phosphate and soda ash solution may be used as a cleanser.

Corrosion in turbine lubricating oil system

Initial corrosion can occur to the pipework, gear casing, sumps, oil tanks etc., during construction. This corrosion if it is not prevented or removed can lead to serious problems—imagine rust flakes falling into the lubricating oil due to vibration then being circulated around bearings and gears.

During construction, control of the environment, use of primary protective coatings (to be subsequently removed in some cases) etc.

is essential. Cleaning with hot air blast, pickling with acid to remove rust and mill scale, washing, drying, applying a protective paint and then sealing the ends is a typical example for pipework.

When the system is erected on board it would be thoroughly flushed through with a hot light mineral oil containing anti-oxidant and rust inhibitor. This would then be replaced with usual service oil.

Even after all the precautions outlined above have been taken corrosion can still occur during normal turbine operation. In tanks and gear cases there will be areas not continuously coated with oil—these may corrode with damaging results to bearings and gears. Salt laden, damp atmosphere can be drawn into the system which can lead to serious corrosion in the vapour spaces. Methods of protecting against vapour space corrosion are: 1. correct application of epoxy resin oil resistant paint to scrupulously clean metal surfaces. 2. use of antioxidants and rust inhibitor in the oil. 3. vapour phase inhibitors added to the oil—these volatilise in the gear case and form a protective anti-corrosive film on the metal surfaces (these may be the only rust inhibitor added).

Emergency Self-closing Stop Valve

This valve is arranged to close automatically and shut off steam supply to the turbines as a result of:

1. Loss of lubricating oil pressure.
2. Turbine overspeed.
3. Excessive axial rotor travel.
4. Low condenser vacuum.

The valve is closed manually by:

1. Handwheel.
2. Emergency hand control.
3. Deck manual control.

Modern practice is to include the valve in a common chest enclosing manoeuvring valve and astern master valve. As a general rule the valve is only arranged to cut off ahead steam, astern steam usually being under direct operator control and such steam may be required to assist in emergency stopping.

Referring to Fig. 4.38:

To open the valve rotate the handwheel, the locking supply lever opens the supply valve (sometimes referred to as an inclined plane action) which allows steam pressure to the left of the piston. The piston is an easy fit in its cylinder so steam pressure equalises on both sides of the piston, providing that the steam drain connection to the condenser is shut, the valve then follows the handwheel opening due to steam pressure acting on it.

The valve closes if the drain is opened by the control valve or low vacuum control, either by automatic or hand action, because steam pressure on the left hand side only of the large piston overcomes the opening action due to steam pressure on the smaller valve acting from the right. If main steam flow reverses, the non-return ball valve closes and traps steam on the left of the piston. This action, together with backflow pressure on the back of the valve, closes the valve, *i.e.* action as a non-return boiler stop valve. The emergency control system is shown in Fig. 4.39.

The steam control valve is kept seated by oil pressure. If this pressure is released or reduced due to pump stoppage, governor overspeed action or hand emergency control valve opening, the steam control valve opens and connects steam pressure to drain. Low

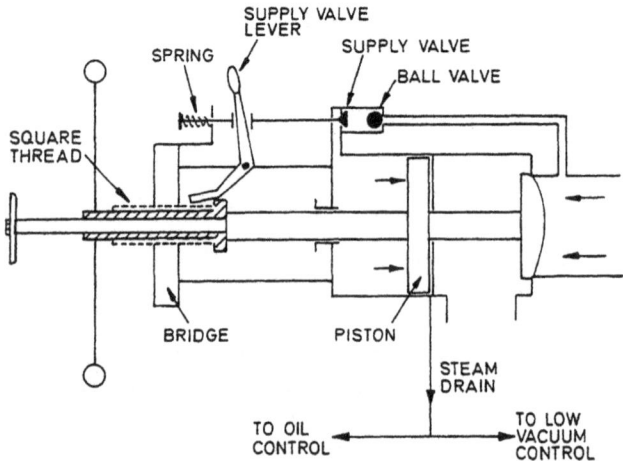

Fig. 4.38 EMERGENCY STOP VALVE (SELF-CLOSING)

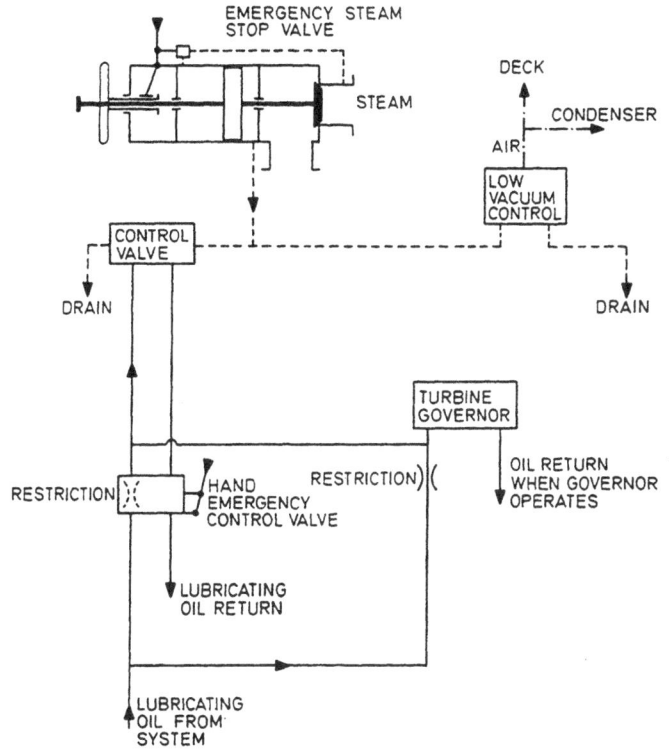

Fig. 4.39 EMERGENCY CONTROL SYSTEM

vacuum operation, emergency hand action from deck or condenser pressure directly, also opens a drain connection. In any of these cases the emergency stop valve will close.

Overspeed and Excessive Axial Movement Trip Mechanism

Land turbines etc. normally driving electrical units operate at varying loads and numerous types of governors are available to control speed right through the load range, an emergency overspeed trip is also provided. There are three main methods of governing: 1. Throttle governing. 2. Nozzle control governing. 3. By-pass governing. Most governors have a centrifugal operating principle functioning through an oil servo or relay to the throttle or to a camshaft which controls nozzle valves.

Marine turbines for main propulsion are almost at constant load and the governor is therefore more of an emergency governor or overspeed trip. However, for automated plants speed control of the turbine is required and various means are adopted *e.g.* a speed sensing governor. This together with the Aspinall emergency trip governor, with excessive rotor travel trip will be considered. The units are built into the control system.

Referring to Fig. 4.40:

The oil supply passes through a restricting orifice to index bush E and then through annulus G and ports L into annulus D around the catch spindle.

The governor valve A is offset from the centre of the shaft so that it is unbalanced when rotating. It is held in position by the spring. When the turbine overspeeds the unbalanced weight overcomes the spring pressure and the valve moves out until catch spindle C can move into the notch in the valve and hold the valve open. In this position the oil in annular space D is allowed to discharge through ports J, and, since the discharge area is about ten times the supply area at the restricting orifice, the oil pressure is released and the steam valve shuts. The tripping speed can be varied by nut B. To reset the valve, the catch spindle is pulled out by knob K allowing the valve to spring back.

When axial movement of the rotor occurs the control bush N moves with it but the index bush E remains still so that if movement is excessive oil escapes from annulus G to the discharge annulus H and oil pressure is once more released and the engine is stopped. The adjustment for this mechanism is in the form of an eccentric F, rotation of which moves the index bush E axially.

Indicator I shows the fore and aft position of the rotor.

Fig. 4.40 has been somewhat simplified from original drawings, in this case sectioning has necessarily been retained otherwise the sketch of the Aspinal mechanism would lose any resemblance to the actual component. This sketch finally presented will need practice to produce quickly under examination conditions but it should be remembered that it is illustrating two protective functions.

Also included in Fig. 4.40 is a governor of the eccentric ring type,

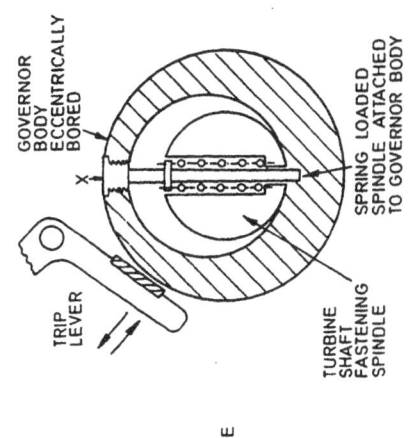

EMERGENCY OVERSPEED
TRIP (GOVERNOR)

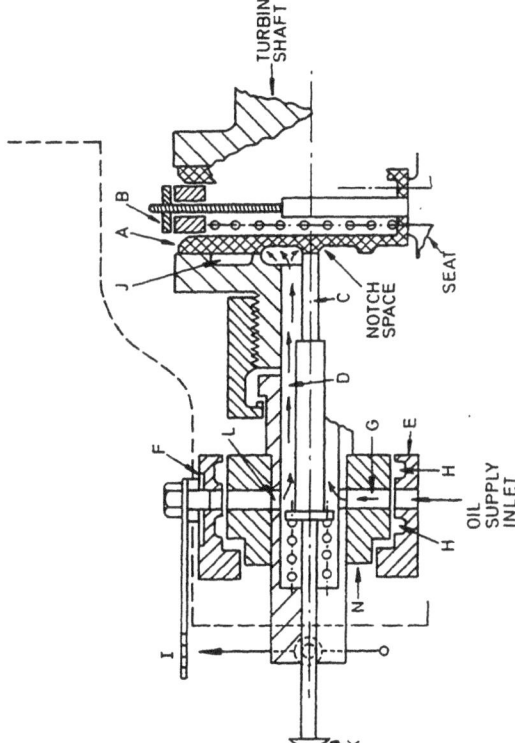

'ASPINALL' OVERSPEED AND EXCESS
AXIAL MOVEMENT TRIP MECHANISM

Fig. 4.40

this would be suitable as an emergency trip for an electrical turbo-generator. The trip is screwed on to the end of the turbine shaft. The body ring is eccentric to the spindle so that the resulting centrifugal force is opposed by the compression of the spring. Adjustment is achieved by the screw at X. At normal speed the outside of the ring is coaxial with the turbine spindle. At excessive speed the thicker part of the ring moves out compressing the spring and overcoming its resistance. The trip gear functions to shut off steam supply usually through an oil relay system.

Fig. 4.41 shows in diagrammatic form a speed control governor. It consists of a non-rotating oil supply tube and a rotating assembly, the latter incorporates two cantilevers with ball valves at their free ends.

As speed increases the cantilever arms move radially out at their free ends opening the ball valves and allowing oil to escape to drain. This reduces oil pressure in the supply line which in turn acts upon the servomechanism for the ahead manoeuvring valve which would then close in reducing steam supply.

Low Vacuum Trip
Refer to Fig. 4.42.

With the advent of all steel low pressure turbine casings this fitting

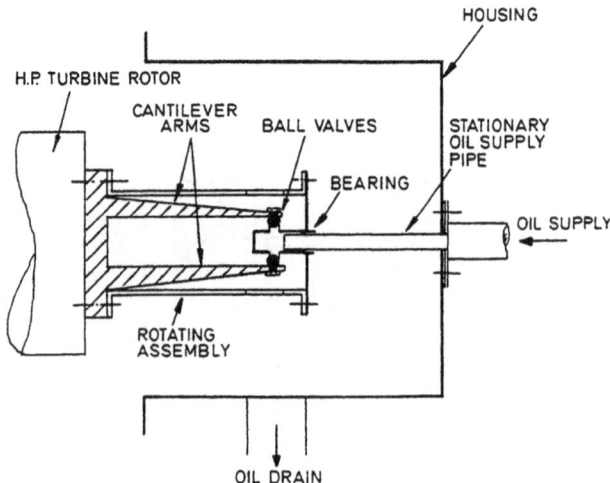

Fig. 4.41 GOVERNOR FOR SPEED CONTROL

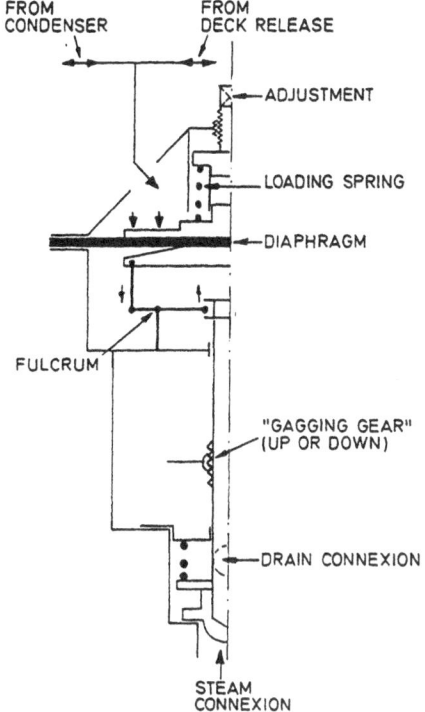

Fig. 4.42 LOW VACUUM TRIP

is often omitted to simplify the control system. Considering the sketch given.

Increase of pressure on the diaphragm top, by decrease of vacuum or hand release from deck, causes down movement. By the linkage the valve opens, so connecting steam pressure to drain. This action closes the emergency stop valve. The gear works equally well by releasing oil pressure and functioning on the control valve in a similar way to the other oil operated trips.

Astern Turbine Safeguards

When operating under normal ahead running conditions the astern turbine in the l.p. casing is in a vacuum environment. It is important

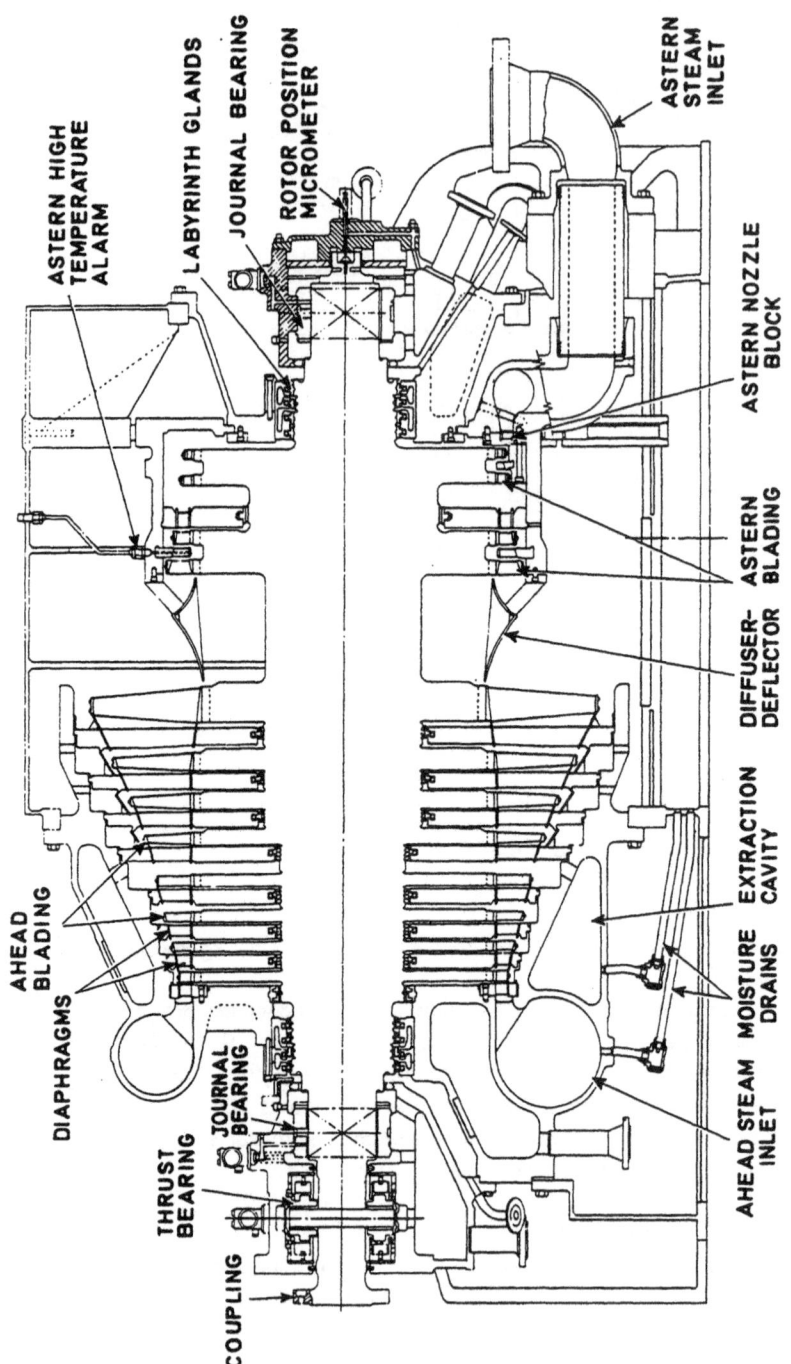

Fig. 4.43 MODERN L.P. TURBINE (30,000 KW SET)

that no steam leakage past the astern manoeuvring valve occurs, as this would leave the astern nozzles at high velocity, enter the astern turbine and due to pumping, friction and windage cause overheating with possible damage. It has been estimated that 1% leakage at full power may reduce the ahead power by 5% or more.

A double shut off arrangement is often used as a precaution. This consists of a astern 'guard valve' between the astern manoeuvring valve and turbine. A drain fitted to the steam line between these two valves could give indication of steam leakage past the manoeuvring valve. But if this drain was left open and the guard valve leaked, air would enter the l.p. turbine casing.

A detector for steam leakage past the manoeuvring valve can compare astern nozzle box pressure with condenser pressure (there should be no pressure differential running ahead). This combined with a temperature probe in the astern turbine casing (see Fig. 4.43) to give indication of overheating should give sufficient protection.

TEST EXAMPLES

Class 2

1.   Describe how to prepare a set of geared turbines for manoeuvring out of port. How is a turbine engined ship put astern? What is the normal full astern power achieved expressed as a percentage of full ahead power?

2.   Sketch and describe an emergency steam stop valve as fitted to turbine installations.

3.   Sketch and describe the gland steam arrangement for a turbine set. State which valves are open and shut under any given circumstances.

4.   State what is meant by end tightened blading. Sketch and describe an adjustable thrust and explain clearly how adjustment is made and detail the working clearances utilised.

5.   Sketch and describe the packing as used on an impulse turbine diaphragm plate. What materials are used in its construction?

6.   Sketch and describe a combined turbine thrust and journal bearing for an impulse turbine. How is expansion of the rotor allowed for?

7.   What is meant by double casing construction when applied to

turbines? What advantages and disadvantages are there with this form of construction?

Class 1

1. Sketch and describe a bearing, thrust and labyrinth gland for the forward end of a turbine.

2. Sketch and describe a double casing turbine. What means are provided to allow for expansion? Detail the materials used in construction. State the advantages claimed for such a turbine.

3. Sketch and describe the various types of labyrinth packing as used in turbines. Explain how steam leakage is kept to a minimum and how air leakage is prevented.

4. Sketch and describe a turbine sliding foot showing clearly where the foot is situated on the turbine. Describe in detail the warming through of a turbine in readiness for sea. Pay particular attention to the vacuum carried and the clearances to be checked throughout the time interval.

5. Describe a governor system suitable for use with a cross compound turbine set. How is overspeed protection arranged?

6. Sketch and describe a section of end tightened turbine blading. Explain fully its construction and how the clearances should be adjusted after turbine overhaul.

N.B. For more recent examination questions refer to miscellaneous examples.

CHAPTER 5

FUNDAMENTALS OF CONTROL

GENERAL

Automation

The automation of equipment has been given many confusing definitions. *Full* automation can be considered to be satisfied by:

1. Centralisation of *all* instruments and recorders, for the *whole* ship, to *one* control station. This station would be situated in *any* convenient position and would include the control functions as well as recordings.

2. The recording system would utilise an electronic data logger in place of conventional instrument types with associated manual writing of results. The readings would be frequently scanned and recorded at reasonable intervals on an electric strip printer or typewriter. Any deviation from the individual required values would set off an appropriate alarm and the fault location would be ascertained at the control station using, for example, mimic boards of diagrammatic essential circuits using coloured lines, with recording points indicated by small pilot lights.

3. Adjustment and control would be automatic, utilising a computer in conjunction with the data logger so allowing changes in controlled conditions to be programmed, in the correct sequence and between the required limits, so that correction signals could be sent to the control equipment which would then function to correct *as the change was occurring*. Fail-safe would be incorporated into all control system components.

Such full automation would give much more efficient operation, would give better and more easily handled records, allow less watch keeping personnel, etc.

The *whole* ship would be under the control of one watchkeeper officer responsible for *all* aspects such as navigation, propulsion, refrigeration, etc., based at the control station. This officer could be assisted by two ratings, one a technician or craftsman and the other an operative. A small group of officers and ratings would be available for routine maintenance, inspection, cleaning, etc., on day work.

Utilisation

Such automation as described above may be regarded for marine practice as too advanced at the moment. If automation is replacement, in total, of man by more efficient automatic devices for supervision and control of machinery, then this is definitely true although ships are in service at present with many of the aspects described. Ship's engine rooms are really on the threshold of automation. A good measure of the degree of automation is the redundancy of operatives created. The use of centralised control rooms and data logging has led to much recent publicity. This is in no sense full automation, it is merely the grouping of instruments and controls and the use of more sophisticated 'office equipment' to eliminate manual viewing and logging. Likewise bridge control of engines is only *one* of many applications of control engineering, probably of lesser importance in many types of ship.

Full automation as described, whereby a ship may do a voyage without engineers, is still a fairly long way into the future.

Surprisingly the plant itself can be the reliability problem as control and computer control, on shore in particular, is extremely reliable, this is not yet true at sea. As plant sizes increase, reliability becomes even more difficult for these main units. Experience of equipment, staff training, cost analysis, etc., requires consideration as *full* automation is approached but for the present the *degree* of automation will vary. Automatic control in a simple sense has *always* been utilised, for example, safety valves, feed regulators, etc., leading to more developed systems such as auto-steering, combustion control, etc. Modern systems are a natural development on more sophisticated lines.

Economy

In every case the centralised use of instrumentation and controls can result in a saving in crew numbers. This is not necessarily the primary object as for example the main function of unmanned engine rooms at night is to allow increased day workers which should result in better maintenance and cheaper repair bills. It is however obvious that a saving in manpower can be achieved particularly for skilled engineers.

The machinery will run under more efficient conditions assuming a fairly comprehensive and properly applied system. This is perhaps more particularly important in steam turbine practice where fractional gains over the years have been steadily increasing efficiency. Efficient logging will allow better records, future planning and maintenance in a planned manner.

The detailed analysis of costs requires a full work study, but that economy is achieved is beyond dispute.

Safety

Essential requirements for unattended machinery spaces (designated u.m.s.), *i.e.* particularly unmanned engine rooms during the night could be summarised thus:

1. Bridge control of propulsion machinery.

The bridge watchkeeper must be able to take emergency engine control action. Control and instrumentation must be as simple as possible.

2. Centralised control and instruments are required in machinery space.

Engineers may be called to the machinery space in emergency and controls must be easily reached and fully comprehensive.

3. Automatic fire detection system.

Alarm and detection system must operate very rapidly. Numerous well sited and quick response detectors (sensors) must be fitted.

4. Fire extinguishing system.

In addition to conventional hand extinguishers a control fire station remote from the machinery space is essential. The station must give control of emergency pumps, generators, valves, ventilators, extinguishing media, etc.

5. Alarm system.

A comprehensive machinery alarm system must be provided for control and accommodation areas.

6. Automatic bilge high level fluid alarms and pumping units.

Sensing devices in bilges with alarms and hand or automatic pump cut in devices must be provided.

7. Automatic start emergency generator.

Such a generator is best connected to separate emergency bus bars. The primary function is to give protection from electrical blackout conditions.

8. Local hand control of essential machinery.
9. Adequate settling tank storage capacity.
10. Regular testing and maintenance of instrumentation.

Display

Essentially this aspect consists of centralised instrumentation in an air conditioned instrument and control room. Improved visual, audible and observation techniques are required. The data logger is perhaps the latest device requiring attention. This requires some knowledge of electronics. Components are virtually all electric-electronic (solid state devices working under air conditioned states are preferred) to fit in with standard equipment. Faults will be located by mimic board type diagnosis and replacement of printed card components rather than on the job repair will be essential.

In selecting alarm circuits great care must be taken in the preference choice utilised. Important circuits should be fitted with distinctive alarm indications and a quick and easy position location. Less important circuits can be fitted with a secondary importance alarm and isolating-locating system. The provision of too many alarms, not easily discriminated from each other, can cause confusion. Similar remarks apply to remote control room gauge boards where only really essential measurements should be *frequently* scanned.

The control room itself requires careful design with reference to comfort, lack of lighting glare, selective positioning of instruments for rapid viewing, correct placing of on-off and position and variable quantity indicators, improved instrument indication techniques, rapid control fault location and replacement, etc.

Various types of indicators and recorders are in use, for example: lights, dial gauges with pointer, colour strip movements, magnetic tapes, cathode ray (or G.M.) tubes, counters, charts, etc.

References are usually set on a pinboard and supply voltage stabilisation is usually necessary. Solid state devices show a high reliability rate compared to the older thermionic 'radio' valve circuits especially after a run in period. Typewriters and printers usually require the most maintenance and attention.

Alarm scanning and data logging, terminology

Scanning

The scanner normally covers up to about 200 points at the rate of about two points per second.

Fig. 5.1 shows a multipoint installation for a turbine plant. The indicator, switches, mimic diagram and legend would be situated in the control room. For scanning and recording the switching is performed automatically by means of a small motor driven unit and the signals received would then be amplified, passed to an analogue to digital converter and thence to alarm, print out and display units.

Measurement

All analogue imputs are amplified from the low voltages produced by the instruments. This signal as a voltage representation of the measured value is translated in the analogue-to-digital converter to a numerical code form.

Display

The code signal is transferred to a strip printer or electric typewriter, printing is selected for the various points at preset intervals, varying from virtually continuous for certain points, to reasonably long time intervals for others.

A second function is to compare digitally the analogue inputs with preset limit switches or pins in a patchboard and have lights on mimic diagrams to indicate alarms, in addition the excess deviation readings are presented on a separate alarm printer.

Programme

This is a predetermined scanning routine. Print-out is timed by the special digital clock.

Equipment

Consists of solid-state silicon components on logic boards as printed circuits. Relays are hermetically sealed relay type on plug-in cards. Test board and replacement cards are provided for fault detection and replacement. Data loggers are sectional framework construction, *i.e.* modules.

Analogue Representation

Where the measured quantity is converted into another physical

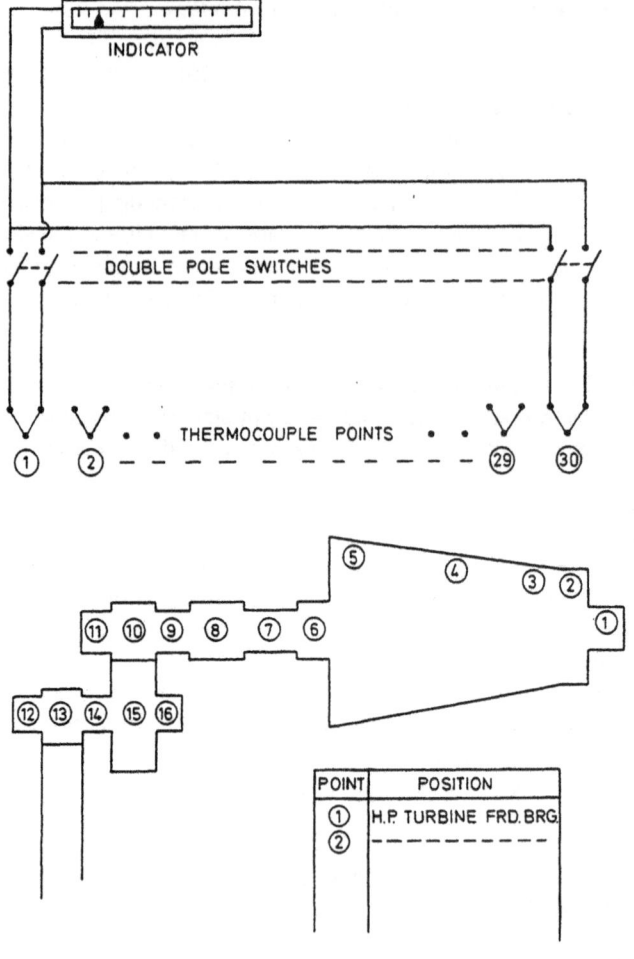

Fig. 5.1

quantity *in a continuous way.* For example temperature converted into d.c. voltage by a thermocouple. Voltage is analogue of temperature. Useful for short term presentation, *e.g.* manoeuvring, raising steam, etc.

Digital Representation
    Where the measured quantity is represented by repeated individual

increments *at given intervals*. For example a revolution counter which trips to alter the reading each engine revolution. Useful for long term presentation, *e.g.* full away watchkeeping readings.

Advantages of Data Logging

1. Reduces staff and number of instruments.
2. Provides fairly continuous observation and fault alarm indication.
3. Provides accurate and regular operational data records.
4. Increases plant efficiency due to close operational margins.

Sensor

Really a *telemetering* device but will be defined here. A device which by utilisation of a physical property gives indication of condition of plant variable. For example the instrument for observing pressure may be a pressure gauge in which the sensing device is the bourdon tube which is then a direct positional indicator. The working principle is mechanical strain.

Transducer

Another mainly *telemetering* device. No appreciable load may· be placed on the sensing device by the indicator or inaccuracies will result. The transducer converts the small sensing signal into a readily amplified output, usually in a different form. For example, mechanical movement to electrical output, electrical input to mechanical output, etc. An illustration of the transducer principle is given in Fig. 5.4 (given later) where mechanical movement is converted to a pneumatic output signal, amplification can be arranged with a pneumatic relay.

Chopper

A transistor multi-vibrator circuit to rapidly open and close the circuit to give alternating current from direct current prior to amplification and rectification. This procedure is usually necessary as d.c. amplifiers suffer from drift when voltage variations at supply are present, hence a.c. amplification is used.

Amplifier

The amplifier (often called a relay in pneumatics) is to step up the sensing low power signal to a high power actuator element. Standard electrical practice at one time utilised cross flux excited d.c. machines (amplidynes) for heavy current and valve amplifiers for light current. Modern communications utilises transistors and silicon controlled rectifiers (thyristors).

CONTROL THEORY

This section is the *basic* requirement to the subject of control. The principles involved are relatively simple yet they have in general been ignored by engineers over the years. A gap has existed in that the measuring instrumentation was understood and the broad functional working of the control loop itself was appreciated but the actual 'guts' of the individual controller was not functionally examined for operating theory.

Control theory is presented here totally by illustration of hydraulic or pneumatic techniques. Electric-electronic devices work on identical principles of operation only differing in component design.

The approach presented is deliberately brief and simple.

Terminology

Correct terminology for automatic control is given in British Standards 1523 Section 2, 1960.

In the author's opinion the complexity of the definitions requires some simplification, the following is an interpretation:

Automatic open loop control system:

Is one in which the control action is independent of the output. An example is a sootblowing system—the output could be the cleanliness of the tubes but this is not used in any way to control the sootblowing action.

Closed loop control system:

Is one in which the control action is dependent upon the output. The system may be manually or automatically controlled. Fig. 5.2 shows the basic elements in a closed loop control system.

The measured value of the output is being fed back to the controller which compares this value with the desired value for the controlled condition and produces an output to alter the controlled

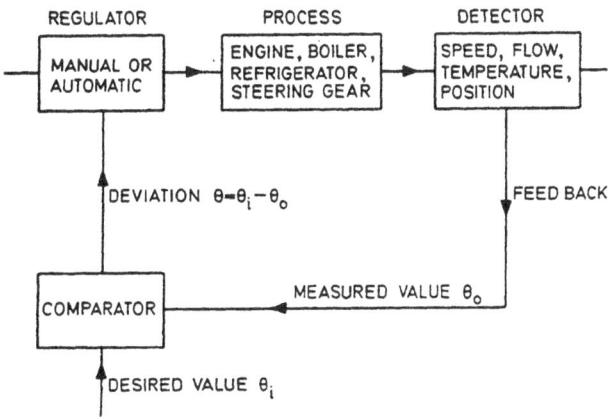

Fig. 5.2 CLOSED LOOP CONTROL SYSTEM

condition if there is any deviation between the values.

Measured Value; actual value of the controlled condition (symbol θ_0).

Desired Value; the value of the controlled condition that the operator desires to obtain. Examples, 2 rev/s, 25 degrees of helm, 55 bar, $-5°$ C, etc. (symbol θ_i).

Set Value; is the value of the controlled condition to which the controller is set—this should normally be the desired value and for simplicity no distinction will be made between them.

Deviation (or error); is the difference between measured and desired values (symbol θ). Hence $\theta = \theta_i - \theta_0$. This signal, probably converted into some suitable form such as voltage to hydraulic output or voltage to pneumatic output, etc., would be used to instigate corrective action—object to reduce the error to zero.

Offset; is sustained deviation.

Feedback; is the property of a closed loop control system which permits the output to be compared with the input to the system. Feedback will increase accuracy and reduce sensitivity.

CONTROL ACTIONS

Three basic actions will be described: (i) Proportional; (ii) Integral; (iii) Derivative. An analysis of their working actions is given by distance-time graphs rather than mathematics because they give a clear pictorial representation. The slope of a distance-time graph is velocity; an inclined straight line is constant velocity as the slope is constant, a curve of increasing slope represents acceleration, a curve of decreasing slope represents deceleration.

Consider briefly a human control loop:

A man regulates a water inlet valve to maintain gauge level in a tank which has outflow demand. He is told the level required (desired value), will see the level (measured value), after a change he will compare his desired and measured values and decide a course of action (the correcting action), finally there is amplification and relay so that his muscles can operate the valve (correcting element).

Proportional control

Fig. 5.3 shows a simple proportional action control loop which appears to have the same effect as the human control loop. Consider say an increase of demand which causes the level to fall and the float lever to open the valve. Valve opening, by the simple lever principle, is proportional to fall of float.

Consider a demand change:

A point will be reached when the valve will be at a new position where supply equals demand and the level will remain constant. The level will not return to the original since if it did the supply would be back to where it was before the demand changed. Proportional control will arrest the change and hold it steady but at a different point from the desired value. This difference between new value of level and desired value of level is called offset. This is the shortcoming of proportional control.

Offset can be reduced by increasing the sensitivity of the system (*i.e. narrowing the proportional band*) e.g. in Fig. 5.3 move the fulcrum to the right, *i.e.* small float travel then causes big valve travel so that the level would not alter appreciably as it would be arrested due to the large amount of entering water. There is a limit to sensitivity increase because when the valve travel is large the inertia of the

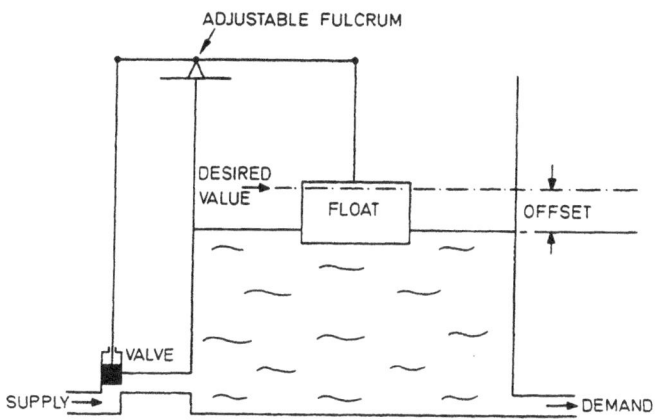

Fig. 5.3 SIMPLE PROPORTIONAL CONTROL LOOP

fluid medium itself would resist such rapid changes and hunting and instability would occur. Decrease of sensitivity *widens the proportional band.*

Proportional band is usually expressed as a percentage (often 'throttling per cent' on controllers). It is a measure of the gain of the controller, the narrower the band (*i.e.* smaller the percentage) the higher the gain and the higher the sensitivity. For example a pneumatic temperature controller reading 400–600° C, *i.e.* range 200° C and output signal 1.2 to 2 bar, *i.e.* range 0.8 bar, then the controller has a gain of unity (proportional band 100% if 200° C variation causes 0.8 bar variation. If 100° C variation causes 0.8 bar variation then gain is two (proportional band 50%). Proportional band is defined as that variation of the measured variable, expressed as a percentage of the range of the measuring instrument, required to give 100% variation in the output signal. Proportional band is usually (not always) adjustable at the controller.

A pneumatic type of proportional controller is shown diagrammatically simplified in Fig. 5.4. A set value of pressure Ps is established in one bellows and the measured value of pressure Pm is fed into the opposing bellows (the measured value of pressure could be proportional to some measured variable such as temperature, flow, etc.). Any difference in these two pressures causes movement of

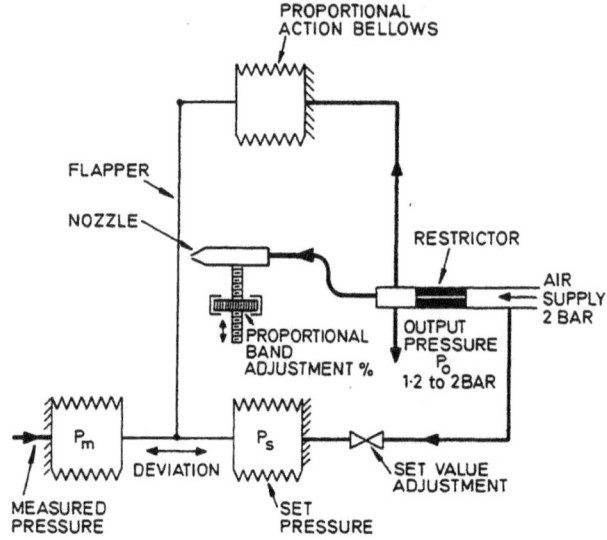

Fig. 5.4 PNEUMATIC PROPORTIONAL CONTROLLER

the lower end of the flapper, alteration in air flow out of the nozzle and hence variation in output pressure Po to the controlled valve.

If the upper end of the flapper was fixed, *i.e.* no proportional action bellows, then a slight deviation would cause output pressure Po to go from one extreme of its range to the other. This is simple proportional control [output pressure change Po and deviation (Pm − Ps)] with a very narrow proportional band width and high gain (gain = controller output change/deviation). Moving the nozzle down relative to the flapper increases the sensitivity and gain, and further narrows the proportional band width.

With output pressure Po acting in the proportional action bellows the top end of the flapper will always move in the opposite direction to the lower end, this reduces the sensitivity and widens the proportional band. When adjusting the controller to the plant the object would be to have minimum offset with stability, *i.e.* no hunting. Commencing with maximum proportional band setting (200%) move set value control away from and back to the desired setting (step input) and note the effect on the controlled variable, using step reductions of proportional band and upon each reduction a step input, a point will be reached when oscillations of the controlled variable do not cease, a slight increase in proportional band setting to eliminate the oscillations gives optimum setting.

Proportional plus integral control (P + I)

In the human control loop previously considered offset would not occur, *i.e.* the operator would bring the level back to the desired value each time after arresting the level. That is he has applied a re-set action. Integral action is aimed to do this re-set, *i.e.* P + I gives an action to arrest the change *and then* restore by reset to the *desired* value irrespective of load. Integral action will always be occurring whilst deviation exists.

Fig. 5.5 shows a diagrammatic illustration of the principle. Note the use of terms like valve positioner, summing unit, two term controller. Fig. 5.6 illustrates integral action signal. Note that whenever the variable is away from the desired value (set value) the integral action is always moving to correct. If offset is acceptable over the range of the variable then a simple proportional controller is acceptable.

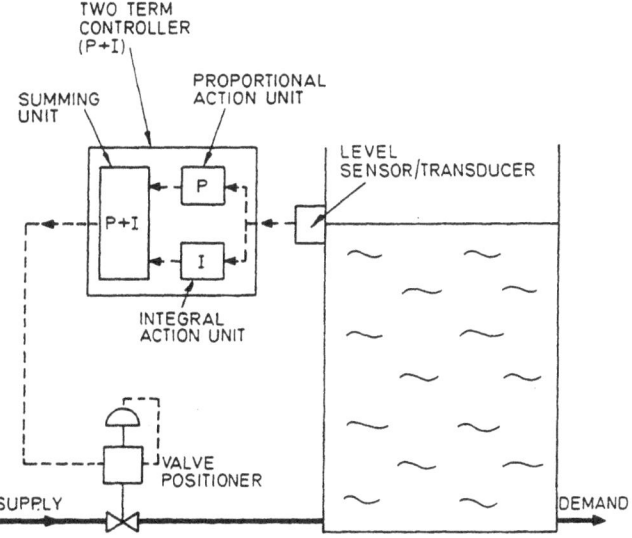

Fig. 5.5 PROPORTIONAL PLUS INTEGRAL CONTROL LOOP

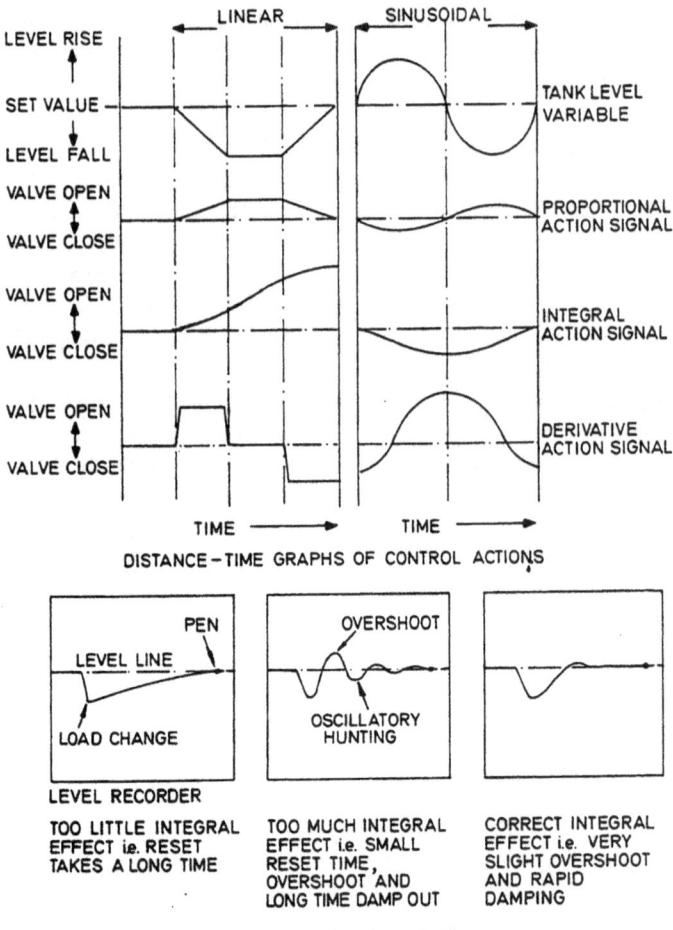

DISTANCE – TIME GRAPHS OF CONTROL ACTIONS

LEVEL RECORDER

| TOO LITTLE INTEGRAL EFFECT i.e. RESET TAKES A LONG TIME | TOO MUCH INTEGRAL EFFECT i.e. SMALL RESET TIME, OVERSHOOT AND LONG TIME DAMP OUT | CORRECT INTEGRAL EFFECT i.e. VERY SLIGHT OVERSHOOT AND RAPID DAMPING |

INTEGRAL ACTION TIME

Fig. 5.6

As another illustration of offset consider a mass oscillating on a spring. This is proportional action. The restoring force will always bring the mass to an equilibrium position *for that mass*. Alter the mass (load) and a different equilibrium position applies—this is so for all different masses so tried. Desired value can only be achieved for one known mass value equilibrium position.

Controllers are usually arranged with an adjustment to vary the integral action time (often expressed as 'Reset time' on controllers). Integral action is always related to proportional control action. Integral action time, for *constant magnitude error*, may be defined as the time required by the integral action to alter the controller output by an amount equal to the proportional action.

For example, if a deviation exists which causes a proportional correcting signal of say 0.3 bar then integral action starts. If it takes one minute for this action to add or subtract another 0.3 bar to the correcting signal then the integrator has an integral action time of one minute (or one repeat per minute). The *shorter* the time setting the *greater* the integral action. Too great an integral effect will cause overshoot past the desired value.

Proportional plus Integral plus
Derivative Control (P + I + D)

In the human control loop previously considered overshoot would not occur, *i.e.* the operator would *not*, whilst adjusting for offset, go on altering the valve right up till offset was gone. The operator would start easing down the valve adjustment rate as the desired value was approached. Integral action must eliminate offset with a minimum of overshoot, this can normally be arranged satisfactorily with a two term controller, *i.e.* P + I. However, a third term, derivative control action may be added as a damping action to reduce overshoot. Derivative control action is anticipatory. If deviation is varying the rate of change of deviation can be used to estimate what the final value of the measured variable is likely to be. If a control signal can be used based on the rate of change of deviation then the controlled condition can be overcorrected and so the likely deviation reduced. Rapid change of deviation will give a large overshoot whilst a slow change will give a small overshoot. When the variable is moving the derivative control action signal will set up a correction signal which is proportional to the velocity of the variable.

Fig. 5.6 illustrates derivative action signal. Note that derivative action opposes the motion of the variable regardless of the desired value.

Derivative action time is usually adjustable at the controller. Derivative action time, *for constant magnitude error*, may be defined as the time required for the proportional action to be increased by an

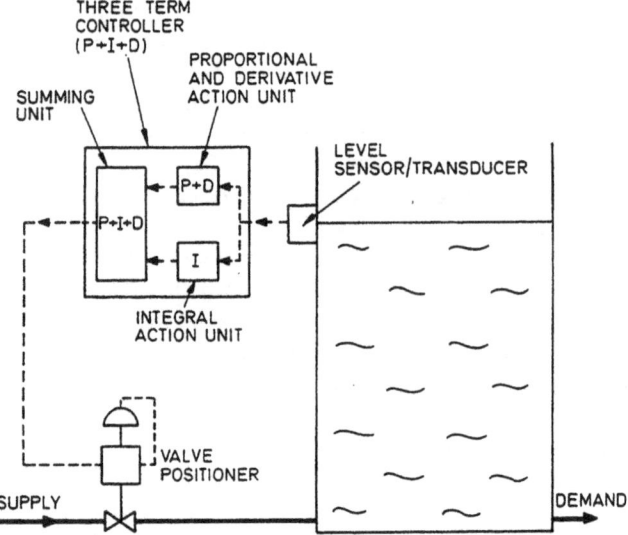

Fig. 5.7 PROPORTIONAL + INTEGRAL + DERIVATIVE CONTROL LOOP

amount equal to the magnitude of the derivative action. The *longer* the time setting the *greater* the derivative action, *i.e.* the more sensitive the derivative control.

Control Actions

Again briefly refer to Fig. 5.6 which shows on the first sketch two deviation effect signals for fluid level in a tank. For proportional action signal note on the second sketch that it is opposite to the deviation, the relative heights of sketch one and two depend on the proportional factor between correction and deviation. For the integral action signal then for the value at any time on the third sketch think of the area developed at that instant on the first sketch applied on the opposite side of the line, and to a suitable scale factor. For derivative action signal then the value of the signal on the fourth sketch is the change of slope of the first sketch, again opposite side of the line, *i.e.* slope only changes at four points on sketch one and at such points the derivative effect is acting almost instantaneously.

Summary

(P) *Proportional control:* action of a controller whose output signal is proportional to the deviation.

i.e. Correction signal $\propto$ deviation.

(I) *Integral control:* action of a controller whose output signal changes at a rate which is proportional to the deviation.

i.e. Velocity of correction signal $\propto$ deviation.

Object: To reduce offset to zero.

(D) *Derivative control:* action of a controller whose output signal is proportional to the rate at which the deviation is changing.

i.e. Correction signal $\propto$ velocity of deviation.

Object: Gives quicker response and better damping.

(P) Single term controller.

(P + I) or (P + D) Two term controller.

(P + I + D) Three term controller.

Pneumatic controller (P + I + D)

Fig. 5.8 shows in diagrammatic form a three term controller. Set value control and proportional band adjustment have been omitted for simplicity (see Fig. 5.4). Often, controller manufacturers produce a standard three term controller and the installer can adjust for type of control action necessary, *i.e.* either single, two or three terms as required.

Proportional only: Integral and derivative action controls shut. Pressure P_0 is approximately proportional to movement of flapper, in relation to nozzle, by the deviation.

Approximate sizes are, restrictor about 0.2 mm bore, nozzle about 0.75 mm bore and flapper travel at nozzle about 0.075 mm. To ensure exact proportionality and linearity the effective flapper travel is reduced to near 0.025 mm, giving less sensitivity and wider proportional band, with negative feedback on flapper due to inner bellows and pressure P_0 acting on it. Note: Whichever way the bottom of the flapper is moved, by the deviation, if the top is moved in the opposite sense negative feedback, if in the same sense positive feedback.

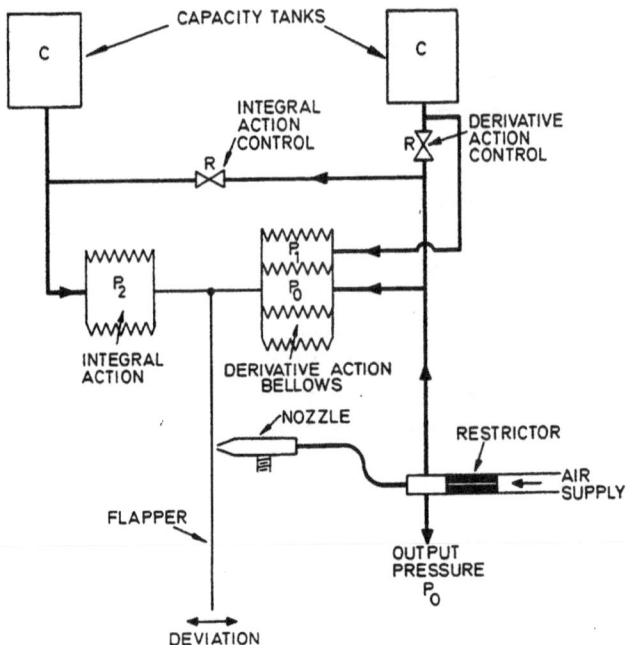

Fig. 5.8 PNEUMATIC P + I + D CONTROLLER

Integral added (P + I)

This is applied by adding positive feedback with pressure P_2 acting on the integral action bellows. Integral action time is the product of the capacity C and the resistance of the integral action control R, *i.e.* RC (note the similarity with electrical circuits). Increasing R by closing in the integral action control increases integral action time.

Derivative added (P + D)

This is applied with further negative feedback with pressure P_1 acting on the derivative action bellows. Derivative action time is the product of the capacity C and the derivative action control resistance R. Increasing R by closing in the derivative action control increases derivative action time.

Note: Integral action is very rarely applied on its own. Derivative action is never applied on its own.

CONTROL PRACTICE

Marine control systems are either pneumatic, hydraulic or electronic, or a combination of these. The electronics could act as the nervous system with the pneumatics or hydraulics supplying the muscle. The advantages of the systems will now be examined and it will be left as an exercise for the reader to list the disadvantages.

Pneumatic system advantages:
1. Less expensive than electronic or hydraulic systems.
2. Leakages are not dangerous.
3. No heat generation, hence no ventilation required.
4. Reliable.
5. Not very susceptible to variations in ships power supply.
6. Simple and safe.

Electronic system advantages:
1. Fewer moving parts hence less lubrication and wear.
2. Low power consumption.
3. System is either on or off, with pneumatic or hydraulic systems if they develop a leak, or dirt enters the system, they become sluggish.
4. Small and adaptable.
5. Very quick response.

Hydraulic system advantages:
1. Nearly instant response as fluid is virtually incompressible.
2. Can readily provide any type of motion such as reciprocating or rotary.
3. Accurate position control.
4. High amplification of power.

Data logger
    The term data logger is loosely used nowadays to describe a broad range of electronic systems that automatically collect and process data, some control and supervise the plant hence some data loggers would better be described as on-line computers.
    Fig. 5.9 shows a very simple data logging system, the elements indicated perform the following functions:

1. *Sensor-transducers.* These are detecting conditions and changes

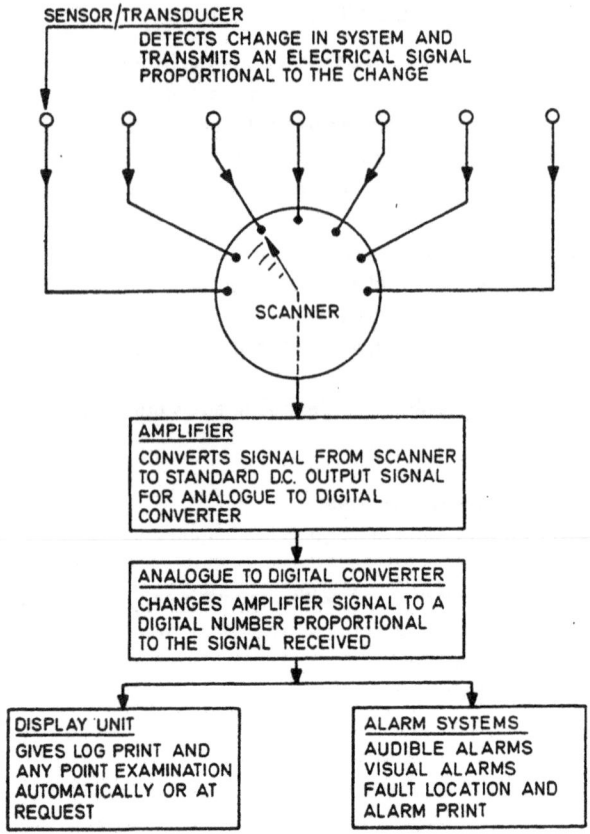

SENSOR/TRANSDUCER
DETECTS CHANGE IN SYSTEM AND
TRANSMITS AN ELECTRICAL SIGNAL
PROPORTIONAL TO THE CHANGE

SCANNER

AMPLIFIER
CONVERTS SIGNAL FROM SCANNER
TO STANDARD D.C. OUTPUT SIGNAL
FOR ANALOGUE TO DIGITAL
CONVERTER

ANALOGUE TO DIGITAL CONVERTER
CHANGES AMPLIFIER SIGNAL TO A
DIGITAL NUMBER PROPORTIONAL
TO THE SIGNAL RECEIVED

DISPLAY UNIT
GIVES LOG PRINT AND
ANY POINT EXAMINATION
AUTOMATICALLY OR AT
REQUEST

ALARM SYSTEMS
AUDIBLE ALARMS
VISUAL ALARMS
FAULT LOCATION AND
ALARM PRINT

Fig. 5.9 SIMPLE DATA LOGGER SYSTEM

in the plant under control such as pressure, temperature, flow, level, speed, power, position, and are converting the signals received into proportional d.c. electric outputs. The term transducer means conversion from one form of signal to another, such as pressure to electrical output, or temperature to pressure output, and so on, but in the case of the data logger system generally the output signals are electric.

2. *Scanner.* This receives the d.c. outputs from the sensor-transducers, which are analogues of the physical functions being measured, into its channels. There may be from 10 to 1000 different

channels dealt with in rotation generally, but it can be equipped with random access facility.

3. *Analogue-digital converter.* This would incorporate the amplifier, the voltage signal received would be amplified and converted to frequency so that the signal is now in a suitable form for digital measurement and display.

An on-line computer would have a programme stored within its memory and it would receive the digital signals from the analogue to digital converter and compare these with its programme, if these disagree then the computer would instigate corrective action. An automatic watch-keeping system for the engine room would have in addition to the foregoing, a shut down protection of the main engine with an over-riding facility in the event of unsafe manoeuvring conditions, automatic change over to standby pumps and emergency generator, automatic bilge pumping and trend monitoring, this gives early indication of possible malfunctions.

In the event of an alarm limit being exceeded (limits are usually set by pins in a matrix board, but with trend monitoring, alarm limits can be varied automatically with ambient conditions) a klaxon will sound and a flashing window will identify the channel. The klaxon can be silenced by pressing an alarm acknowledged button, but the window will remain illuminated until the fault is remedied. The alarm condition and time of clearance will be automatically recorded.

All recorded data has time of recording associated with it and means are provided to enable the time to be adjusted to ships time, alarm print-out is often in red, normal in black. Most modern data logger systems have digital display of measured values whereas older types gave analogue display, such as voltmeters whose pointers traversed scales marked off in temperature, pressure, level, *etc.* Digital display is clearer, more accurate and reliable and by using frequency as the analogue of voltage there is greater freedom from drift.

Solid state devices and printed circuits are extensively used in the data logger system, these increase the reliability, simplify maintenance, are most robust, withstand vibration better and are cheaper to produce.

Some Typical Control Loops in Use

Modern steamships now utilise a high degree of control. The

number and complexity of loops has increased rapidly and as a guide some well proven systems will be described.

Sootblowing control system

These systems may be electrically or pneumatically controlled, Fig. 5.10 shows a typical pneumatic system incorporating, in sequence, retractable blowers and then rotary blowers. The blowing medium may be steam or air. Steam would be supplied to a header arranged with automatic drainage via steam traps, warming up is accomplished by incorporating time delay in the control system.

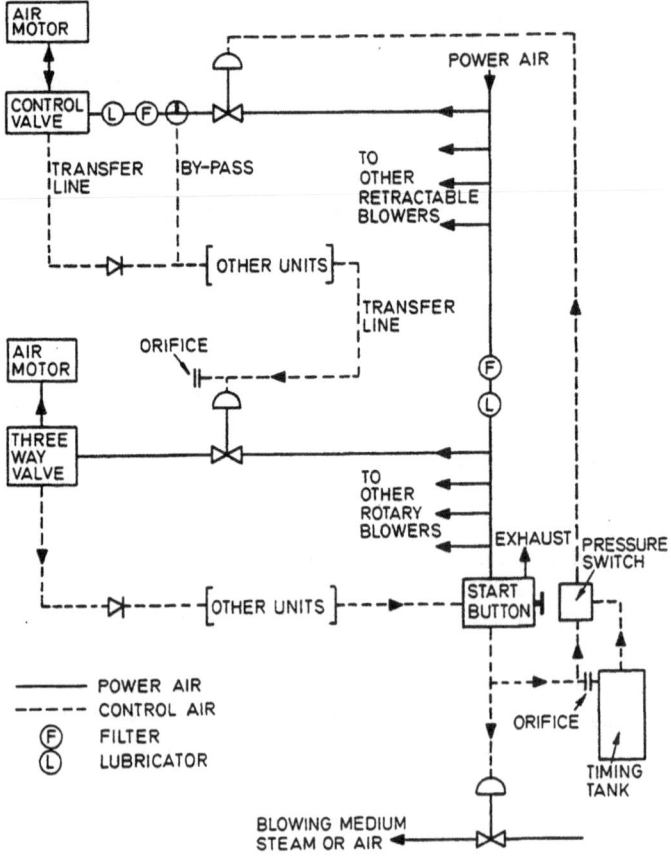

Fig. 5.10 SOOT BLOWER CONTROL

Power air at a pressure of about 8 bar is supplied to the system and the operation sequence is commenced upon depression of the start button. This admits control air to the diaphragm controlled valve for the blowing medium and to the timing tank via the orifice. Pressure will gradually build up in the timing tank thus giving time for warming up of the steam header and connections together with effective drainage. When the air pressure reaches a pre-determined value the pressure switch opens to admit control air to the first diaphragm control valve for blower operation.

Power air now passes to the air motor control valve, picking up lubricant in the process. The blower then moves through full travel out and returns. When fully retracted a trip operates to shut off power air to the blower, the power air can now pass via the motor control valve to the transfer line and become the control air for the next blower diaphragm control valve. This process is repeated until all retractable blowers have been operated (or by-passed if required).

The next blower in the sequence may be of the rotary non-retractable type, transfer air from the last retractable blower becomes the control air for the first rotary blower diaphragm control valve. Power air passes to the three way valve and air motor, the motor rotates the blower element through a certain number of revolutions after which the three way valve cuts off power air to the motor and permits it to pass as control air to the next rotary blower diaphragm control valve.

This is repeated until the last blower in the sequence has operated, then the control air from the last three way valve resets the start button.

With the start button reset the control air to the blowing medium diaphragm control valve and the first diaphragm valve for the retractable blowers goes to exhaust and the valves close. All other diaphragm control valves have orifices fitted through which control air bleeds to atmosphere causing the valves to close and the motor control and three way valves to reset to their starting position.

De-aerator level control (Fig. 5.11)

The tapping points at the top and bottom of the de-aerator water storage tank are led to a differential pressure transmitter whose output air signal (range 1.2 to 2 bar) is proportional to tank level.

The output signal from the differential pressure transmitter is led to

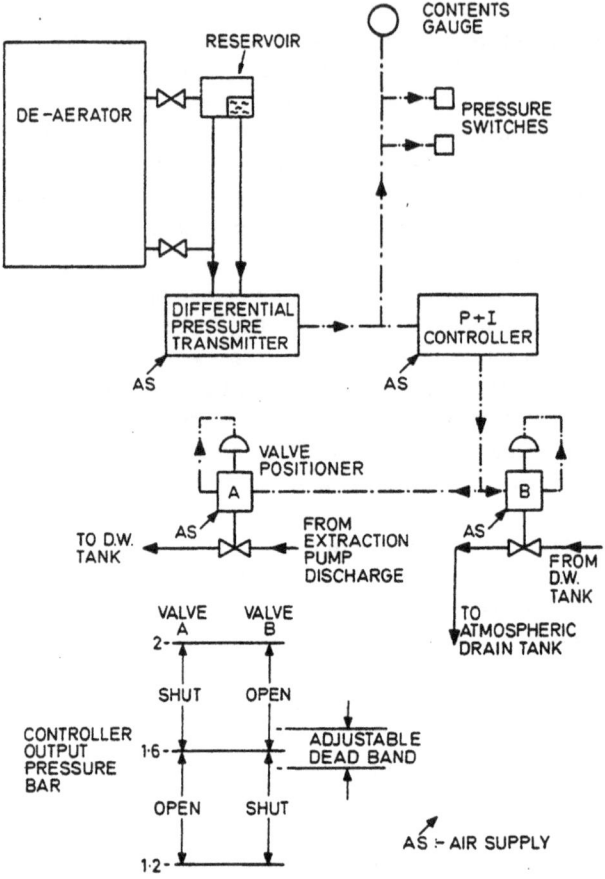

Fig. 5.11 DEAERATOR LEVEL CONTROL (SPLIT RANGE)

a P + I controller where it is compared with the set value, any deviation results in a change in controller output signal to the valve positioners.

Valve positioner A is actuated over controller output range 1.2 to 1.6 bar and valve positioner B 1.6 to 2 bar. This is known as split range (or split level) control, at 1.6 bar both valves would be closed. There is an adjustable dead band over which both valves would be closed, this allows the surge capacity of the de-aerator water storage tank to be used.

Pressure switches actuate audible and visual alarms in the event of high or low water level in the tank, and tank contents are displayed on a gauge. All diaphragm valves are provided with hand jacks for manual operation if required.

This level control system is part of a more general control arrangement in the closed feed system shown in Chapter 6 to which the reader should refer for further information.

Feed water control

1.   Single element control—the element is some form of level control which has already been discussed in Chapter 1.

2.   Two element control—elements are steam flow and level measuring each of which sends out a pneumatic signal proportional to changes in its measured variable. Fig. 5.12.

The steam flow signal passes to the relay which transmits a signal to the valve positioner, via the hand auto unit, which alters the valve position an amount proportional to the steam flow. By correct adjustment of the proportional band in the relay and correct setting up of characteristic in the valve positioner it is possible to match changes in load with feed.

The water level signal is fed into a P + I controller whose set value is adjusted for desired level of boiler water. Any deviation between measured and desired values results in a change in output signal from the controller to the relay. This signal is added to the steam flow signal and is used to correct deviation in level which could occur due to unbalance of boiler feed in and steam flow out, the latter may be due to continuous blow down being put on to the boiler, for example.

3.   Three element control—elements are: steam flow, feed flow and level measuring, each of which sends out a pneumatic signal proportional to changes in its measured variable. Fig. 5.13. The relay compares the two signals from the steam and feed flow transmitters and if these are in the correct ratio, *e.g.* 1 : 1 if no continuous blow down, it transmits a signal to the P + I controller corresponding to the desired level of water in the drum. This signal is compared with the measured value signal being received by the controller from the level transmitter. Any deviation results in a change in controller output signal to the valve positioner.

Considering an increase in steam flow, this would lead to a reduction in drum pressure and a rise in water level (swell effect). When the

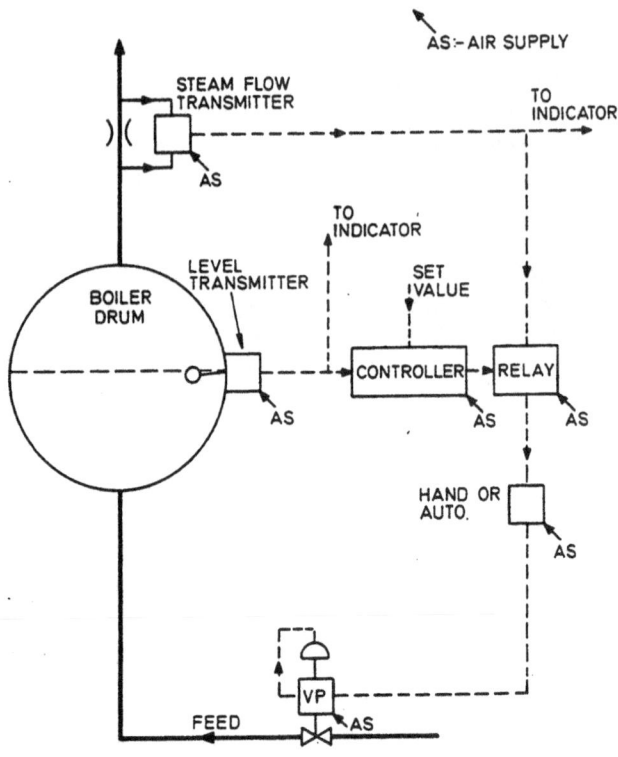

Fig. 5.12 2 ELEMENT FEED WATER CONTROL

steam flow increases the feed valve will be opened to increase feed flow despite the fact that the water level has risen, however, the water level will gradually return to its desired value and a new boiler throughput exists. If the control system were single element, *i.e.* level control only, then upon steam flow increasing and water level rising the feed valve would close in—this at a time when steam demand has increased is satisfactory for low rated boilers, but totally un-satisfactory in highly rated boilers.

Signals from the three elements can be transmitted to indicators and or recorders if desired, also a hand/auto station enables the operator to control the position of the feed valve remotely if required.

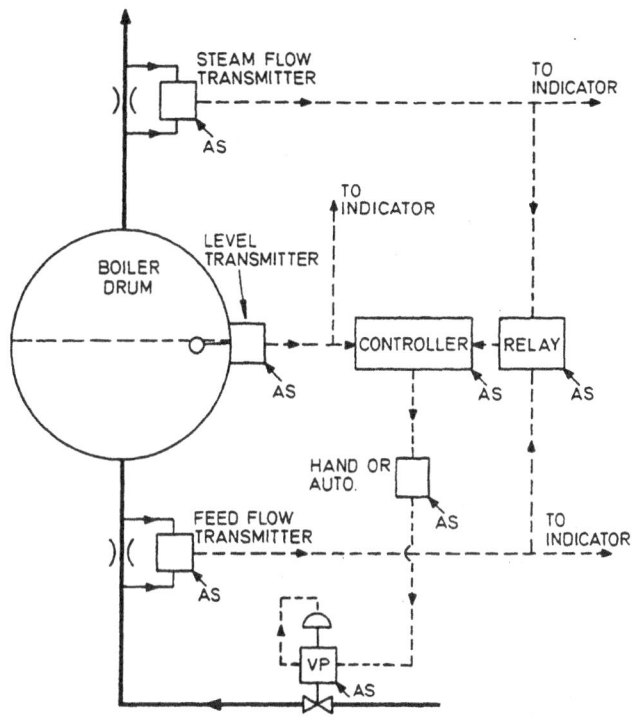

Fig. 5.13 3 ELEMENT FEED WATER CONTROL

Steam temperature control for ESD 111 and
Radiant boilers, etc.

Fig. 5.14 shows a pneumatic control system for final steam temperature. By regulating automatically the amount of steam going through the attemporator (steam cooler) whilst still maintaining full flow through both superheaters, final steam temperature can be controlled.

The steam flow transmitter gives a pneumatic output proportional to steam flow which is fed into an adding relay, the other incoming signal to the adding relay is from the steam temperature transmitter (output proportional to temperature) via the $P + I + D$ steam temperature controller.

Output from the adding relay passes through a hand-auto control

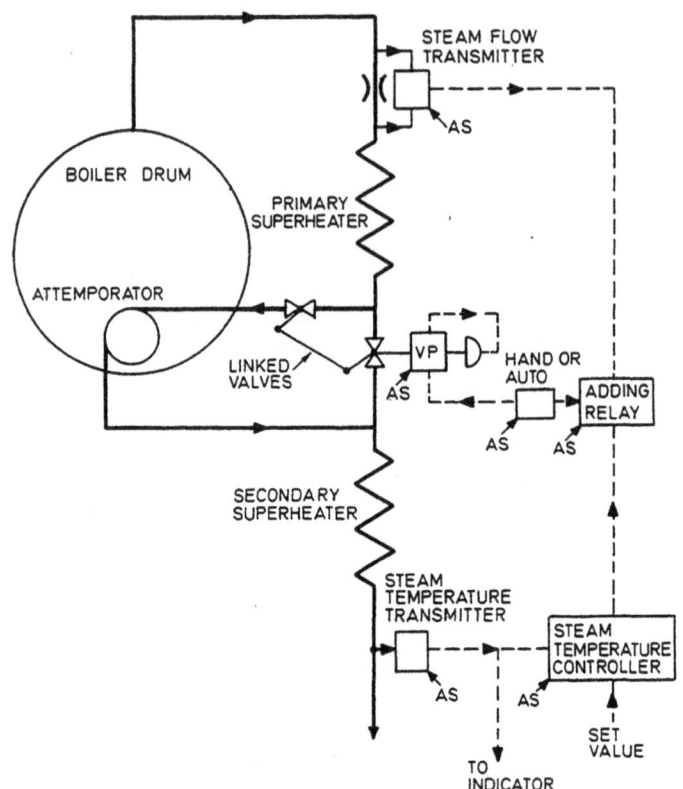

Fig. 5.14 STEAM TEMPERATURE CONTROL

station to the valve positioner. Shown in the diagram are linked control valves with one valve positioner, however, two valves with positioners operating on split range control is an alternative.

By using a two element system better control over steam temperature is achieved during transients, *e.g.* if the demand suddenly increased, increase in steam flow would be accompanied by a fall in steam temperature. However, due to increase in steam flow the control system reduces amount passing through the attemporator tending to keep steam temperature up. The temperature controller, whose set point can be adjusted remotely, would act as a trimming device to restore steam temperature to desired value.

Combustion control

With stricter enforcement of the clean air act coupled with the need for fuel economy and good integrated overall control, a combustion control arrangement for boilers is essential. Automatic combustion control was previously used only at full power, but because of modern demands outlined above the control system must be capable of functioning effectively during rapid load changes.

Fig. 5.15 shows such an arrangement. The steam pressure transmitter sends out a change in pneumatic signal, proportional to the change in steam pressure, to the P + I steam pressure controller. This signal is compared with the set value and any deviation results in

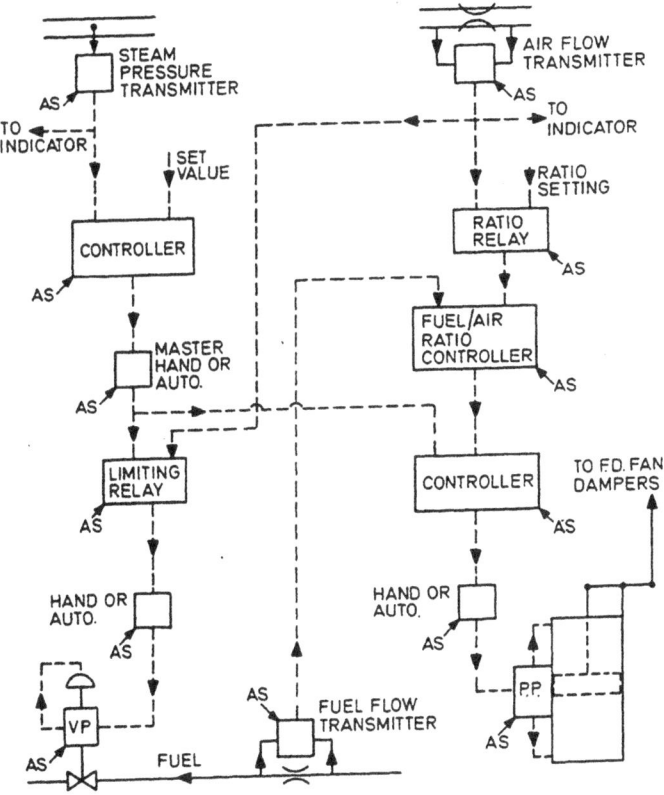

Fig. 5.15 COMBUSTION CONTROL

a controller output signal change which is fed to fuel and air control loops, since a change in fuel supply must be accompanied by a change in air supply.

Measurement of air flow is fed into the ratio and limiting relays. Since the air pressure differential range being measured by the air flow transmitter is small, 0 to 20 m bar, and the full range of transmitter output signal is large, 1.2 to 2 bar, the output signal changes from this transmitter covers only a small part of the design range and hence have to be amplified by the ratio relay.

The gain of the ratio relay can be adjusted remotely and hence its output to the fuel/air ratio controller can provide variations in the fuel/air ratio.

Fuel flow changes result in the fuel flow transmitter producing a pneumatic output proportional to the change which is fed into the fuel/air ratio controller. Output from the fuel/air ratio controller is fed into the P + I controller whose pneumatic signal change goes to the piston positioner (pp) for eventual alteration in air supply.

When steam demand increases, and hence steam pressure falls, more fuel is required. But it is essential that this be preceded by increasing air supply. In order that the fuel/air ratio is not disturbed, the master fuel demand signal from the steam pressure controller and air flow signal are compared in the limiting relay. The purpose of this relay is to delay an increase in the fuel supply until the air supply has been increased.

Conversely when steam demand falls the change in oil flow leads the change in air flow.

The master hand/auto unit enables the operator to act as the steam pressure controller when on hand control.

With some combustion control systems the steam flow measurement from the feed control is fed into a relay along with the signal from the steam pressure transmitter before the steam pressure controller. This arrangement gives better response and closer relationship between steam demand and boiler output. Also, instead of damper control with a constant speed fan for air supply an alternative set up is to control the fan speed.

Wide range fuel oil burners are a very desirable feature with combustion control as they allow manoeuvring without the need to change burners or to incorporate a complex automatic lighting sequence with its attendant problems. Steam atomisation fuel burners having a wide turn down ratio may be used during manoeuvring with change over to pressure atomisers at 'full away.' This gives water economy since steam atomising units can use up to 0.75% of steam output.

Bridge control of turbine machinery
Instrumentation and Alarms

For the bridge console the least instrumentation and alarm the better, alarms should be essentials only and instruments only those vitally necessary. Suggested alarms could be:

1. High salinity. 2. Low feed pump suction pressure. 3. High condenser water level. 4. Low vacuum. 5. Lubricating oil pressure. 6. Tank contents low level. For direct instruments opinion is divided but no more than say another six indications should be necessary. Engine console and alarms would obviously provide full instrumentation. A typical simplified system is given in Fig. 5.16. This system has direct control at steam manoeuvring valves.

This is essentially a combination electro-pneumatic although all pneumatic or all electric can easily be arranged.

The following points with reference to Fig. 5.16 should be noted:

Selector

Bridge or engine room control can be arranged at the selector in the engine room. When one is selected the other is ineffective.

Duplication

Both transmission control systems are normally identical and operation of the one selected gives slave movement of the other.

Programme and timer

The correct sequence of operations is arranged for the various actions. For example, if we were manually getting the plant ready for manoeuvring then some of the following operations, and where required in correct sequence, would have to be done.

1. Steam raised in boiler.
2. Condenser circulating water on.
3. Lubricating oil circulating.
4. Drains open.

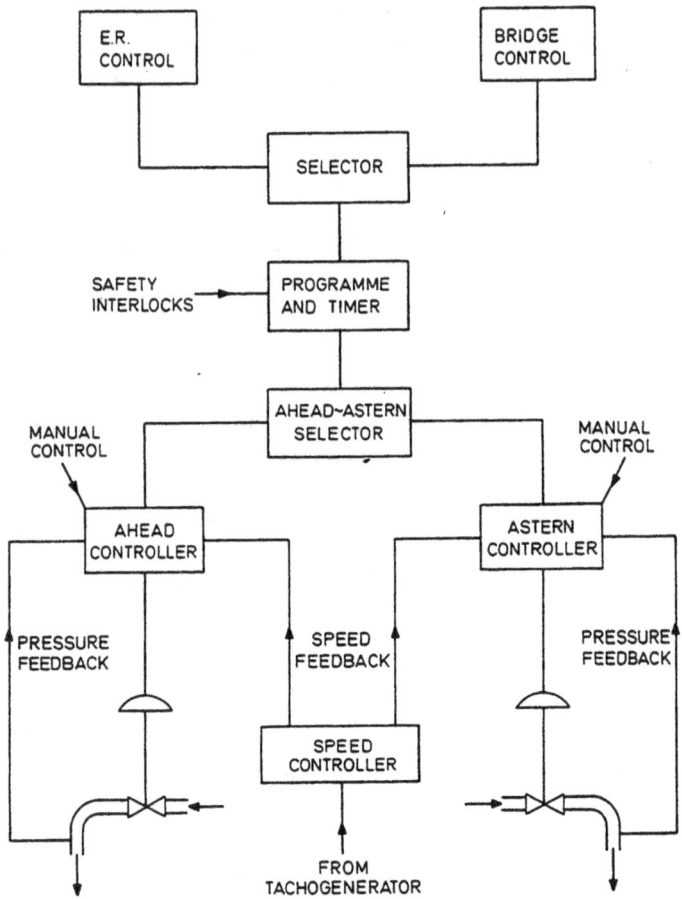

Fig. 5.16 TURBINE CONTROL SYSTEM

5. Turning gear out.
6. Gland steam on, etc.

This then is part of the job of the programme and timer unit, all of the above together with other operations would have to be satisfactorily accomplished before any signal passes to demand opening of a steam supply valve to the turbines.

Timing relays are incorporated to prevent excessive speed

changes, by too rapid signals, which would endanger engines and boilers. The rate of opening of the manoeuvring valve is generally controlled, as sudden opening or closing would give insufficient time for the boiler to adjust itself to the new conditions. An emergency swing from say full ahead to full astern would allow astern braking steam usage under time controlled conditions.

When manoeuvring, a mode switch would be put in the on position. This opens the astern guard valve, shuts off bled steam to heaters, operates main circulating pump at high speed and opens l.p. drains if speed falls below a certain value.

If the turbines are stopped during manoeuvring they would automatically go onto auto-blast, *i.e.* the automatic time delayed opening of the ahead and astern steam valves for short periods—this can be blocked if required.

Protection devices and safety interlocks would be incorporated in the control system some of which could be:

1. Low lubricating oil pressure.
2. Electrical failure.
3. High or low water level in boiler.
4. Condenser circulating water.
5. High condensate level.
6. Low vacuum.
7. Turbine overspeed, axial displacement and vibration.
8. Turning gear in, etc.

Controllers

These would generally be P + I together with trimming relays. The steam pressure feed back gives accurate positioning of the valve for desired pressure and the speed feed back is arranged so that difference in speed between measured and desired values causes an additional trimming signal to the controller. This may be necessary as pressure and speed are not well correlated at low speeds.

With speed feed back incorporated into the system the speed is controlled at a desired value irrespective of changing conditions. Without it frequent changes in control settings would be necessary to maintain steady rev/min.

In the event of emergency direct hand control of the manoeuvring valves can generally be employed. It may also be possible to go from automatic to manual remote control, *i.e.* positioning of the valves by means of their actuators by controlling the change in air, electric or hydraulic signal to them.

Outline Description

The following is a brief description of one type of electronic bridge control for a given large single screw turbine vessel to illustrate the main essentials. Movement of a control lever modifies the output of an attached transmitter (electronic signal 0–10 mA d.c.). The transmitted signal is passed, via override, alarm and cut out units, to the desired flow module which is connected to a time relay and feeds to the controller. The electronic controller compares desired speed with actual speed as detected by a tachometer generator and d.c. amplifier. The correct controller signal is passed to the manoeuvring valve positioner from which a return signal of camshaft position is fed back to the d.c. amplifier, thus giving the command signal to the actuator reversing starter.

The two control levers are independent and do not follow each other, an engine room override of bridge control is supplied. The rate of valve opening is controlled by the actuator so that too rapid valve opening is prevented by a time delay, full normal valve operation shut to open or vice versa occurs in about one minute, this can be reduced to about 20 seconds in emergency by full movement of telegraph from full speed direct to, or through, stop.

A near linear rev/min to control lever position exists. Autoblast refers to the automatic time delay opening of the ahead manoeuvring valve for a short period after a certain length of time stopped—this has an override cut out for close docking, etc.

With regard to overall equipment for automatic control and centralisation then obviously each ship will have an individual requirement varying with degree of automation, ship type, amount of instrumentation, etc.

Steam flow control

1. *Throttle valve control*

This is the simplest method of control. As the valve is closed to reduce steam flow, throttling of the steam takes place, *i.e.* a reduction in pressure with constant enthalpy (*i.e.* constant total heat). Fig. 5.17 illustrates different amounts of throttling and the effect: ab represents available heat energy per kg of steam, as the valve is closed in to

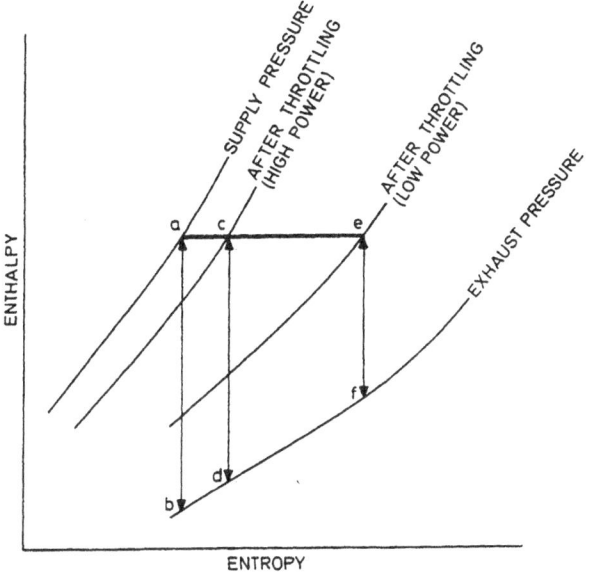

Fig. 5.17

reduce pressure to c or e the heat energy available is reduced to cd and ef respectively. This represents a thermodynamic loss.

2. Nozzle valves

By grouping the nozzles and controlling steam supply to each group, steam flow can be effectively altered without causing the available heat energy per kg of steam to be altered. This gives improved efficiency. Fig. 5.18 shows the bar lift, cam operated nozzle valve control. The bar lifts the nozzle box valves in a predetermined pattern depending upon power required, when it is at F all valves are fully open.

3. Hybrid

This is a combination of throttle valve and nozzle valve control. One nozzle group, about half the total area of the nozzles, is controlled by the throttle valve and the remainder by the nozzle valves—throttle valve combination. This reduces throttling loss at low powers.

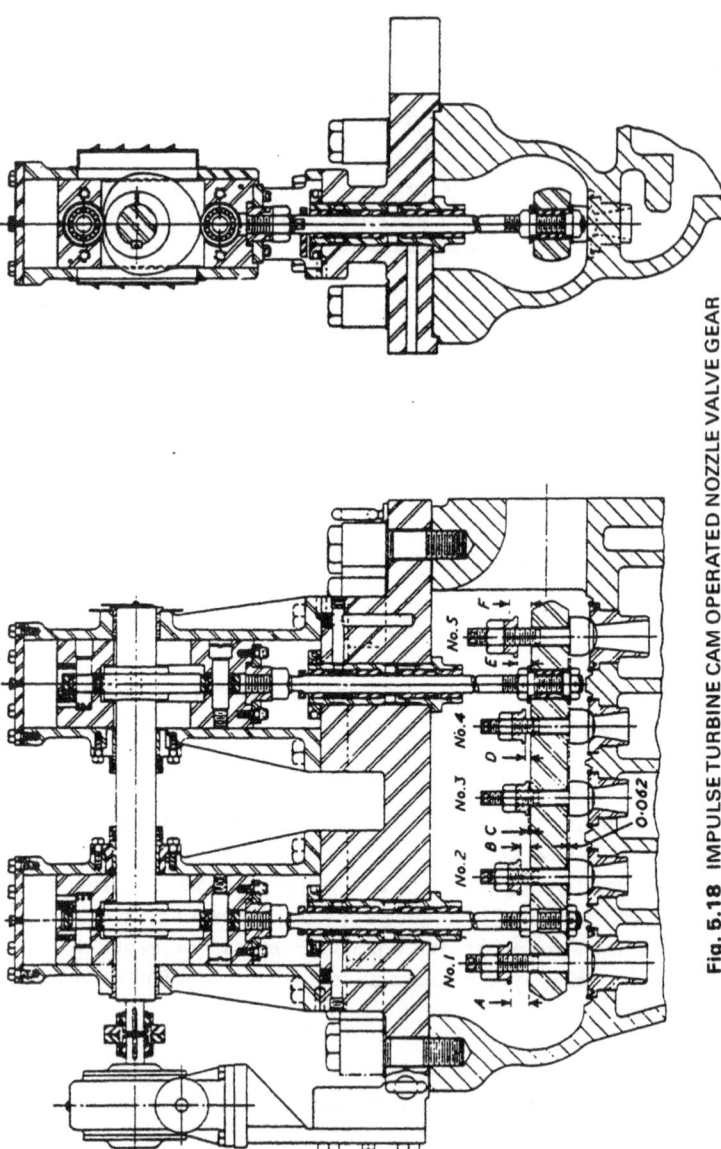

Fig. 5.18 IMPULSE TURBINE CAM OPERATED NOZZLE VALVE GEAR

4. *By-pass valves*

When full pressure is applied to all of the first stage nozzles the steam flow, which is dependent upon pressure drop through the nozzles and total nozzle area, can be increased with by-passing. The by-pass valve takes steam from the first stage and delivers it to a later stage where the nozzle area is large enough to pass the desired steam flow. Hence the power is increased. It would be possible to control steam flow to turbines by varying the boiler pressure, but the thermal lag involved would preclude this, *i.e.* the response to a desired change of steam flow would be too slow.

TEST EXAMPLES

Class 2

1. Describe how liquid level and revolutions per minute may be converted to electrical signals for a data logger, (Electrotechnology Paper).

2. Explain the meaning of the undermentioned terms relating to automatic control, etc.: desired value, error signal, detecting element, feedback, servo motor, reset action.

3. Discuss the problems involved if normal watches were dispensed with and the machinery of a large steamship was controlled from the bridge.

4. Describe any type of pneumatic controller and state its particular function.

5. With reference to control systems explain with the aid of diagrams what is meant by the following: closed loop control, deviation, offset, proportional control.

6. Sketch and describe a two element feed water control system and explain what happens when steam demand rapidly varies.

Class 1

1. With reference to data logging systems, explain the following terms:

(i) sensing device
(ii) scanner
(iii) transducer
(iv) scaling unit   (Electrotechnology paper)

2.   Describe an automatic sootblowing system suitable for a water-tube boiler. Explain the sequence of operation and give a reasoned explanation for the pattern followed.

3.   Describe generally the essential features of a system for remotely controlling the main engine and make specific reference to the arrangements for:
(a) reversing,
(b) changing to load manual control.

4.   With reference to an automatic combustion control system of marine water tube boilers, explain:
(a) the operation of the master controller following variation in steam pressure,
(b) why the pressure drop across the air registers is measured,
(c) how the air fuel ratio is adjusted,
(d) the arrangement adopted to enable continued operation in the event of failure of the mechanism of a fuel-oil flow regulating valve.

5.   In a modern steam plant explain what controls are used for dealing with large differences in water carried in the system between full power and minimum power, where the make-up water passes through a mixed bed de-mineralisation plant.

6.   With reference to the data logger in a centralised control system, explain the advantage of (a) an electric system, (b) a pneumatic system.

7.   Explain the effects of throttling the steam to a turbine with the manoeuvring valve and with the nozzle valves.

CHAPTER 6

FEED SYSTEMS AND AUXILIARIES

That part of the thermodynamic steam cycle which lies between exhaust steam leaving the engine and entry of feed water to the boiler, including all auxiliary equipment and fittings in the line, constitutes the feed system.

As a broad classification this subject can conveniently be split up into three main sections, namely, open feed systems, closed feed systems and miscellaneous related equipment. Each feed system will be described and then all units constituting this system will be considered individually, finally equipment related to feed systems will be considered.

OPEN FEED SYSTEMS

Such feed systems were installed on most steam reciprocating and steam turbine engine plants built before 1945. The plant had the great advantage of simplicity but the need for maximum heat retention in feed water for higher efficiencies, closer control and deaeration for use with water tube boilers in particular, etc., meant that the system was steadily improved until it was totally superseded by the closed feed system.

A typical open feed system circuit is as sketched in Fig. 6.1.

Referring to Fig.6.1.

This system would be suitable for use with low pressure boilers and steam turbines. A typical set of pressures and temperatures to be expected are shown on Fig. 61, for initial steam conditions: Boiler Steam pressure 21 bar; temperature of steam 330° C (115° C of superheat); saturation temperature 215° C.

Note:

1. That the vacuum shown at this sea temperature is very good for

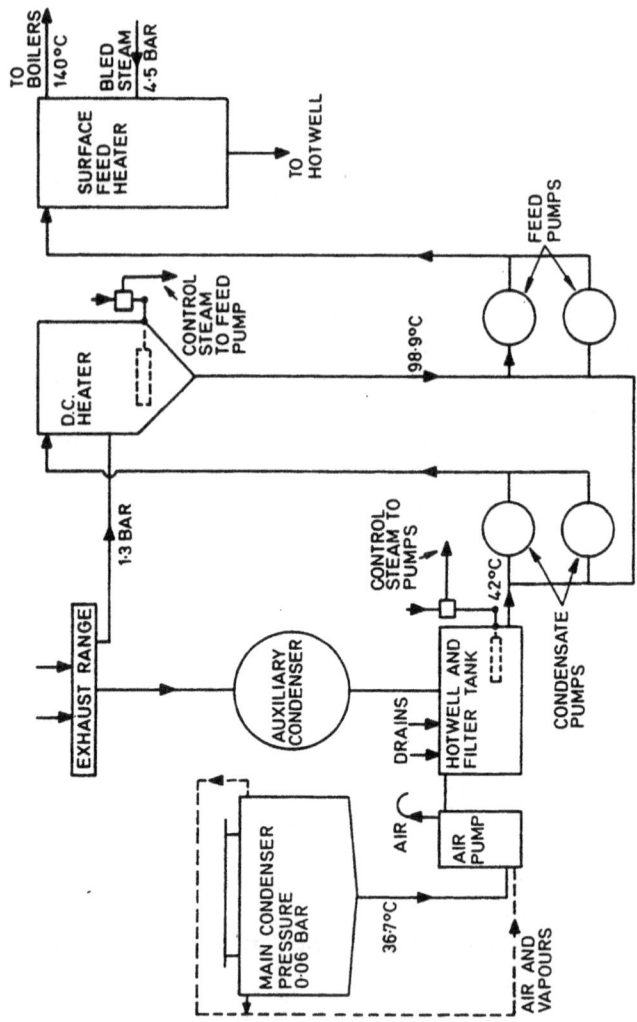

Fig. 6.1 OPEN FEED SYSTEM

such a system, it is based on 760 mm Hg barometer. 2. That the circulating water temperature differential is about 8° C. 3. Undercooling at the condenser is about 1.7° C. 4. The hotwell temperature is very dependent on the amount of heat in live and exhaust steam drains as led back. 5. Only one surface feed heater would normally be used. Water valves and unit valves have been omitted for simplicity.

Referring to the system directly:

The path of the feed water from condenser to boilers can be clearly seen. The air pump has no duplication and speed is only controlled by the hand setting of the steam stop valve. Feed pumps are duplicated and are speed controlled by the water level in the hotwell or in the d.c. heater (float operates a rotatable plug with ports in the steam line). Condensate pumps are duplicated and are speed controlled from the water level in the hotwell.

For manoeuvring, to prevent overheating at the d.c. heater, either (a) d.c. heater steam is shut off only or (b) d.c. heater steam is shut off, condensate pump shut down and feed pump draws directly from the hotwell and delivers directly to the surface heater.

For port use one of the condensate pumps is utilised as a harbour feed pump. The auxiliary condenser supplies to the hotwell and the harbour feed pump delivers to the boiler via the surface feed heater.

Unless water tube boilers are fitted it is not essential to utilise a boiler feed water regulator with the open feed system.

The individual auxiliary units will now be considered.

Air Pumps

A good example of an independent pump as used with an older turbine plant is the Paragon air pump.

Referring to Fig. 6.2.

This pump is twin barrelled, effective stroke about 0.8 m and diameter about 0.5 m. The steam drive is from a conventional steam cylinder and valve gear and due to the fulcrum arrangement both plungers are utilised with about nine double strokes per minute.

Foot, bucket, head and paragon valves are of the circular plate kinghorn type with the paragon chest extending partially around the circumference of both chambers, the buckets are fitted with ebonite rings.

Principle of operation:

On the plunger downstroke air vapours pass above the bucket through the paragon valve chest where the pressure is about 0.04 bar. Near the bottom of the stroke water passes through the bucket

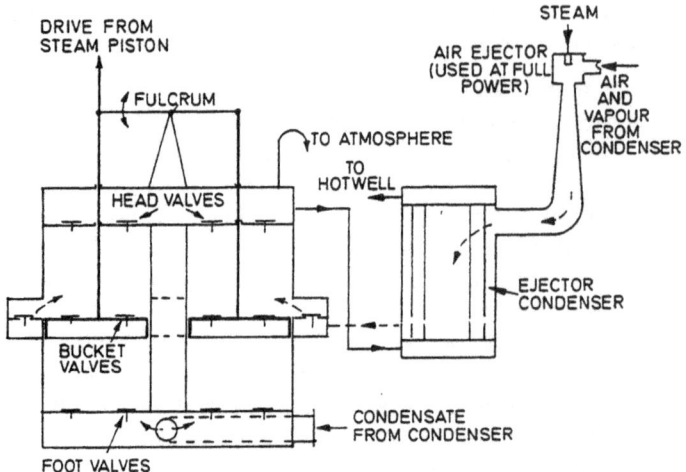

Fig. 6.2 PARAGON AIR PUMP

valves on to the top of the bucket. On the upstroke air and water are discharged above the bucket while water is being drawn through the foot valves. To improve efficiency, air extraction is assisted by using an air ejector. The air is cooled and reduced in volume in the ejector condenser giving up heat to the circulating condensate. The air ejector is not normally used during manoeuvring, as overheating can occur, the pump alone creating quite a satisfactory vacuum.

Separate handling of condensate and air vapour allows full cooling effect on the air vapours to reduce to minimum volume whilst maintaining full heat in the condensate.

Note:

Before the introduction of this pump the Dual air pump was used. This pump was similar in construction. One barrel was used as the wet pump handling condensate and the other as the dry pump handling air vapours, no air ejector was utilised, the wet pump operating temperature being about 8° C above that of the dry pump.

The dry pump was cooled by a closed circuit of fresh water, cooled by a small condenser, any excess water and the air vapours being discharged through a spring loaded valve to below the wet pump head valves.

Function of an Air Pump

Firstly to withdraw air vapours from the condenser to maintain as high a vacuum as possible to reduce back pressure and increase expansion work.

Secondly it is to withdraw the condensate at the highest possible temperature.

It should be noted that the size is governed by air volume (desired vacuum) and the pumps previously described only lift about 12 mm depth of water per stroke on top of the bucket.

Advantages of Independent Pumps:

Independent pumps have the advantage of speed control so that adjustment can be made for higher vacuum and efficiency. Against this, the first cost, maintenance and running cost are relatively high compared to modern arrangements.

Feed Filters

Suction filters are usually incorporated into a combined gravitation feed filter, hotwell and float tank. A common type is as sketched in Fig. 6.3, the operating principle of which can easily be seen. These filters are useful in older steam turbine plants, with steam driven auxiliaries of the reciprocating type for the removal of lubricating oil. In modern feed systems basket type filters would be fitted before the feed pump suctions and after the de-aerator.

Discharge filters, in pairs, are usually of hair felt or fabric enclosed in a pressure shell for pressures to 4.5 bar, metallic filters are preferred for higher pressures.

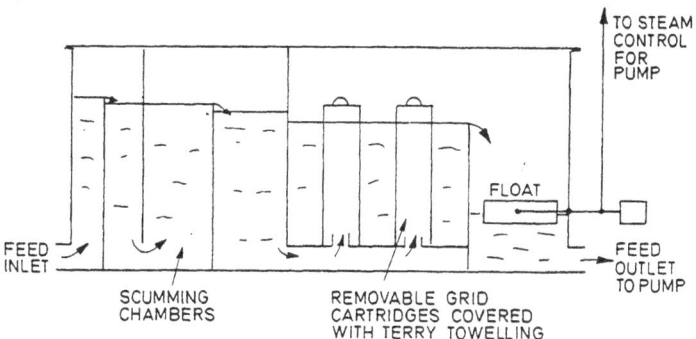

Fig. 6.3 GRAVITATION FEED FILTER, HOTWELL, FLOAT TANK

Feed Heaters

Surface feed heaters have a number of advantages over direct contact heaters, *i.e.* 1. No possible contamination from auxiliary steam. 2. Require less operating attention. 3. Higher feed temperatures can be obtained. As the feed is heated at discharge pressure there is no flash off to steam which can cause much trouble at the suction side of a pump working from a d.c. heater.

The method utilised for feed heating depends very much on the plant design. As much heating as possible, up to about 104° C, is achieved before the feed pump by utilisation of live drains and exhaust steam. The degree of heating the feed pump discharge water depends on the final temperature desired at the boiler.

In general not more than three stages of feed heating would be utilised, *i.e.* two heaters and boiler economiser or three heaters, in marine practice. Some general considerations relating to feed heating will now be considered.

Feed Water Heating Conditions

All of the latent heat and all or part of the sensible heat can be retained within the system when bled steam (or waste steam) heating is used.

With bled steam extraction then allowance must be made for reduced mass flow after extraction point, this is normally done at the design stage.

The advantages of feed heating are:

1. Improved thermal efficiency.
2. Latent heat utilisation.
3. Assists de-aeration.
4. Prevents high temperature differentials at boiler.

Steam should normally be extracted at the lowest possible pressure with sufficient temperature differential for heat flow. The extraction steam pressure is normally arranged to be about 5.5° C above the feed water outlet temperature. In land practice multi-stage cascade heating is used, with say six heaters, which limits the feed temperature rise per heater to 28° C to 39° C but higher differentials are required in marine practice as weight factors normally limit the number of heaters.

Types of Heaters

These are normally classified as low pressure (extraction pump discharge side) or high pressure (feed pump discharge side). Examples of low pressure heaters are direct contact heaters, drain coolers, etc.

Such heaters may be horizontal or vertical, the latter has the advantage of less space requirement for tube nest removal with the overhead crane gear. Heaters are either straight tube or U tubed, tubes of solid drawn steel or brass.

Air vents are provided, preferably situated at the greatest distance possible from the steam inlet as air separation increases with distance. Such vents are usually led to the condenser. Air seriously reduces heat transfer rate in any heat exchanger.

Drainage of heaters is by float traps, drain pumps, loop seal pipes or direct valve control.

The Direct Contact Feed Heater

Referring to Fig. 6.4:

Feed water, from the condensate pump discharge, enters through a spring loaded feed inlet valve. The sudden pressure drop and temperature rise to which the feed water is exposed effectively liberates the air. The water falls through the perforated cylinder and there is intimate mixing of water and steam.

Water level is controlled by a float which controls the speed of the feed pump.

The heater is intended for use with low pressure exhaust steam as the feed outlet temperature is limited so that a satisfactory suction can be maintained at the boiler feed pump. Thus it is usually found that this heater is situated high up in the engine room, a feed outlet temperature of 115° C would require the heater at least 8.3 m above the feed pump.

It will be appreciated that the principle of operation of the d.c. heater is an obvious introduction to the modern de-aerator.

The surface feed heater

Consider the sketch of the surface heater:

The tubes are expanded into the fixed and free (to allow for expansion) tubeplates, access plugs being provided in the outer casing for examination. The temperature of the feed depends on the pressure of the available steam, there is no limitation on the grounds of

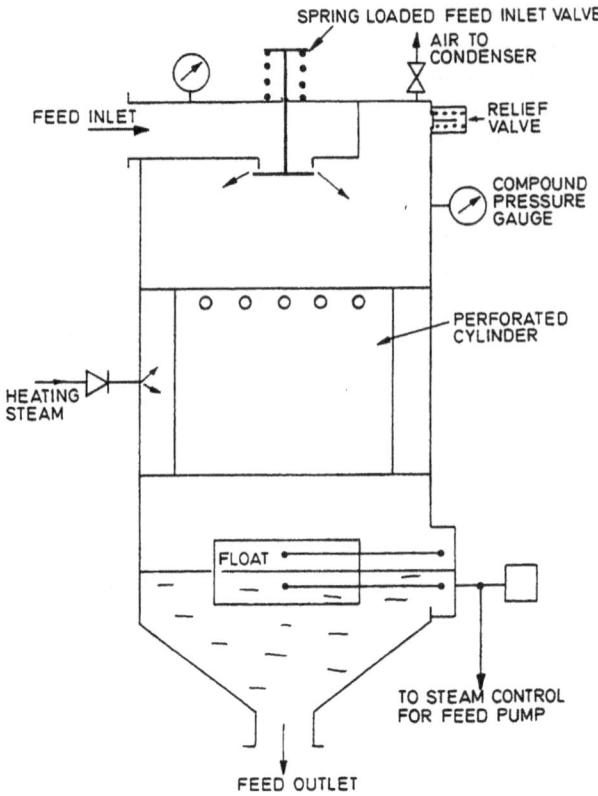

Fig. 6.4 DIRECT CONTACT FEED HEATER

vapourisation as the feed is heated at pump discharge pressure. Exhaust steam at about 1.4 bar gauge would give a feed temperature of about 102° C whilst live steam at 6.3 bar would give a feed temperature near 150° C. The heater should be operated to maintain about $\frac{1}{2}$ glass for most efficent working. The design is multipass to ensure maximum heating surface, baffles are often provided to distribute the steam flow, air release and steam boiling out connections are provided. With the heater in service only the drain requires attention to maintain the correct water level.

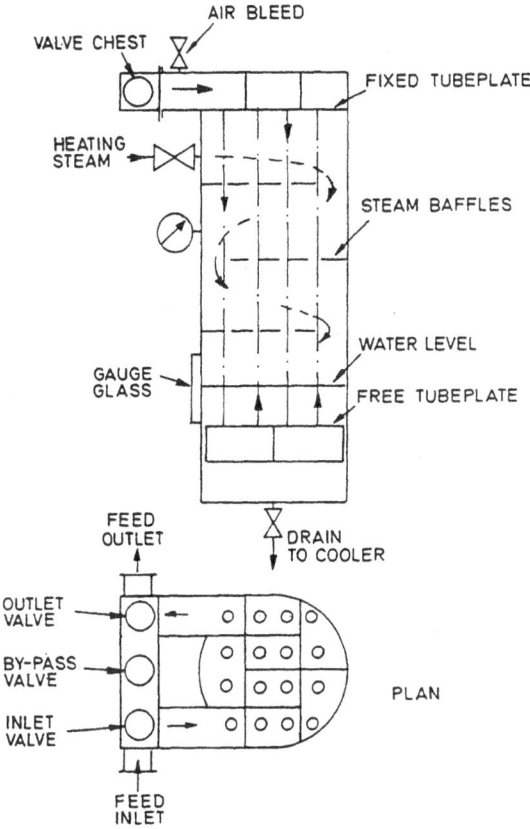

Fig. 6.5 MULTIPASS SURFACE FEED HEATER

CLOSED FEED SYSTEMS

There are many variations on a basic system but a fairly typical closed feed system is sketched in Fig. 6.6.

Referring to Fig. 6.6:

This system would be suitable for use with high pressure water tube boilers and steam turbines. A typical set of pressures and temperatures to be expected are shown in Fig. 6.6.

For initial steam conditions:

Boiler steam pressure 56 bar; temperature of steam 510° C (240° C of superheat); saturation temperature 270° C; feed temperature leaving boiler economiser 200° C.

1st stage heater bled or h.p. exhaust steam 3.2 bar; 2nd stage heater bled steam 7 bar.

Note:

1. The vacuum is based on 760 mm Hg barometer. 2. Undercooling at the condenser is about 0.5° C. 3. Feed heating temperatures and pressures vary considerably from plant to plant depending on design. 4. One factor that influences system conditions greatly is the amount and availability of auxiliary exhaust steam. This will depend on whether turbo or diesel generators are used, whether turbo or electric feed pumps are used, tank heating and general auxiliary demand, etc. 5. This means that the conditions existing in a closed feed system depend vitally on the initial design adopted for heating, bleeding, exhausting, etc. 6. Valves have mainly been omitted for simplicity.

Referring to the system directly:

The path of the feed water from condenser to boilers should be carefully followed. Note that pumps are duplicated, in some cases an electric pump is utilised for full power, the manoeuvring standby pump is steam turbine driven, this again varies from installation to installation. No valves or cocks are shown except the recirculation valve and the screw down non-return valve, this latter valve can be opened in emergency to put a direct suction to the feed tank.

Most of the auxiliaries shown on the sketch will be described individually later. Extraction and feed pumps usually have failure alarm and stand by cut in from a pressure switch. The recirculation valve is usually hand operated and is arranged to be full open when engines are stopped so as to maintain a circulation in units with steam on. The position of this valve can be varied but it is best situated so as to include most units within the recirculation system, the valve is usually shut at full engine power.

The controller on the condenser functions to pass surplus water to the feed tank or draw water depending on the boiler requirements under fluctuating load conditions, this is described later.

The de-aerator is usually situated high up in the engine room, the

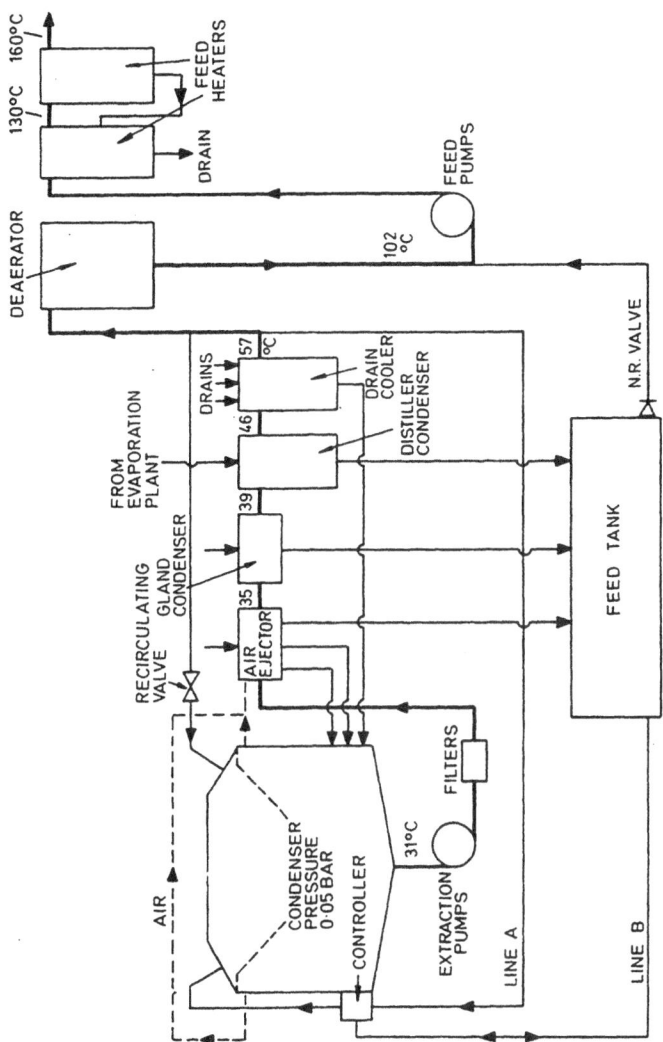

Fig. 6.6 CLOSED FEED SYSTEM (EARLIER TYPE)

level in the associated feed tank is often controlled by the feed rate from the extraction pump discharge. In some installations the de-aerator steam is shut off during manoeuvring but in many cases the capacity of this plant can cope with load fluctuations quite satisfactorily at this unit.

Reserve feed from the feed tank to the condenser level controller passes through the controller to the condenser top, this is usually done to subject the water to condenser de-aeration.

For harbour use a completely separate system is utilised if an auxiliary boiler is fitted. The de-aerator and feed heater system may be utilised from the main plant or the auxiliary plant may employ its own system.

Advantages of the Closed Feed System
May be summarised as follows:

1.   Minimum heat losses and maximum regenerative heat utilisation in the circuit.

2.   Maximum vacuum obtained with air ejector having no moving parts (hence little maintenance) and low weight.

3.   Minimal air leakage into system (positive pressure on all glands).

4.   No feed contamination because of the closed circuit.

5.   Rotary pumps with good efficiency, low weight, reduced maintenance.

Closed Feed Controller
Referring to Fig. 6.7:

Consider the water level falling in the condenser so that the float falls. Water is drawn from the feed tank through line B. The water passes through the supplementary feed valve to line C. This leads water from controller to top of condenser and so into the system, thus the water is heated and further is subject to de-aeration again. Note these lines on the sketch of the closed feed system.

Consider the water level rising in the condenser so that the float rises. Water passes back from the extraction pump discharge due to no demand and flows through line A and Line B to the feed tank.

Between upper and lower travel there is a 50 mm dead travel, with both valves shut, this provides for normal fluctuations without the controller operating.

A relatively modern closed feed system arrangement is shown diagrammatically in Fig. 6.8. It incorporates two stage feed heating (l.p. heater and de-aerator). The l.p. heater takes bled steam from the

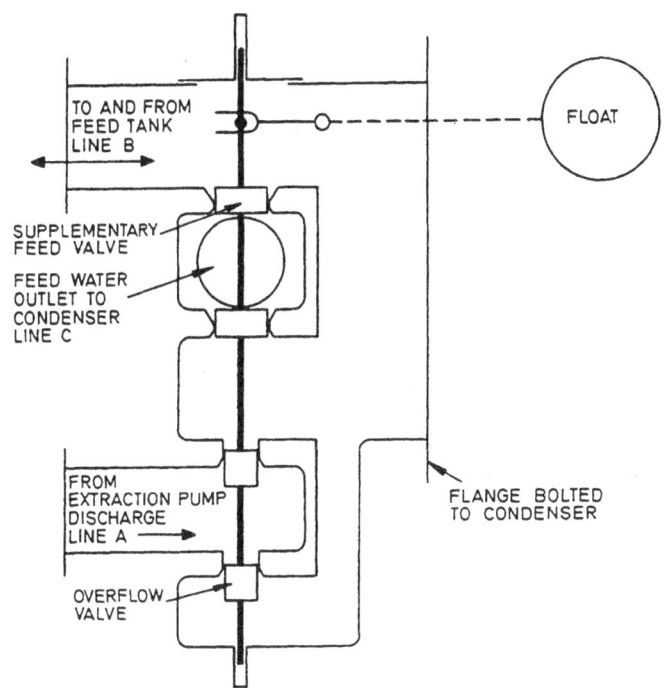

Fig. 6.7 CLOSED FEED CONTROLLER

l.p. turbine and exhaust steam from the first stage bled steam air heater for heating the feed. The de-aerator heating steam comes from the h.p. turbine exhaust (bled) feed pump exhaust, steam to steam generator and the second stage bled steam air heater. Final feed heating is done by the economiser. For simplicity some of the auxiliary circuits have been omitted, but approximate temperatures and pressures at salient points are included since they are often requested in examination questions.

When steaming at maximum rate the amount of water in the boiler will be at a minimum. If the steaming rate is reduced, the quantity of water in the boiler must be increased to re-establish the water level in the drum. To accommodate changes in boiler water requirements of this nature a storage tank containing hot, de-aerated feed water is desirable as the feed water in a modern plant is of high purity and

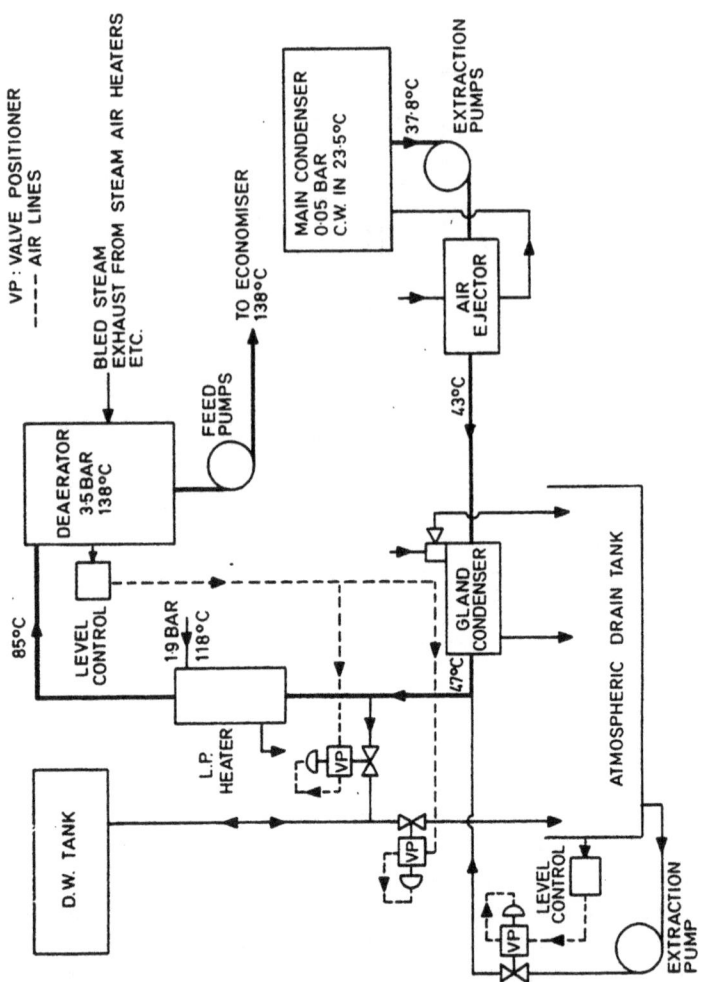

Fig. 6.8 CLOSED FEED SYSTEM

chemically conditioned this would be preferable to transferring feed to and from the system.

The storage tank is usually integral with the de-aerator, see Fig. 6.24, and the water level is controlled over a given range by means of pneumatically operated valves. In the event of the level falling to below a pre-determined value a valve opens and passes feed from the d.w. tank to the atmospheric drain tank. The level in the drains tank will rise and its extraction pump will deliver more feed into the system.

Control over the atmospheric drain tank level is automatic, a pneumatic P + I controller sends a signal to a valve positioner which adjusts the valve throttle on the drain tank extraction pump discharge—high and low level alarms are fitted to this tank level control system.

In this closed feed system only de-aerated feed will enter the boiler, the system will never become 'open.'

An alternative arrangement used on some tankers is depicted in Fig. 6.9 and operates as follows. Again water is stored in the de-aerator storage tank and its level controlled by pneumatically operated diversion and make-up valves. The feed is taken from this

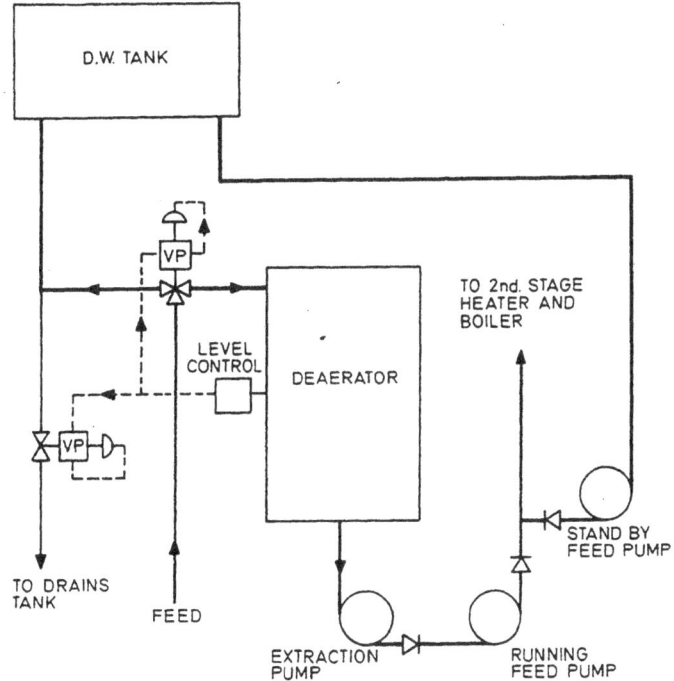

Fig. 6.9

tank by an extraction pump and delivered to the feed pump suction. Normal variations in tank level are controlled as outlined previously. However, in this case, if the de-aerator extraction pump should fail the main condensate is fully diverted to the d.w. tank, the feed supply to the de-aerator being shut off.

The boiler feed pump will lose its suction and its overspeed trip will shut off its steam supply. The standby feed pump (these pumps are interchangeable) which is arranged to take suction direct from the d.w. tank automatically cuts in and the feed is restored to the boiler on 'open' system.

This automatic change over to an open feed system is accompanied by indication to the plant operators who can ascertain and remedy the fault without having complete shut down.

N.B. Level controls have been dealt with in the control chapter.

Feed water re-circulation control

To ensure an adequate supply of cooling medium to condense air ejector steam supply and gland steam, etc. during reduced power operations and when the turbines are warming through, recirculation of the feed back to the main condenser is necessary. The re-circulating valve may be hand operated, remote controlled or automatically controlled.

Fig. 6.10 shows re-circulation control based on the steam pressure in the turbine cross over pipe. When the pressure in this pipe falls to 1.8 bar or below a pneumatic signal passes from the transmitter to the valve positioner and the re-circulating valve opens (two step controller action). Indication that the valve is open or closed is given at the control station where a hand control unit is incorporated for remote operation of the valve.

Alternatively, the re-circulation control can be based upon temperature, Fig. 6.11. When the feed water temperature after the air ejector reaches a pre-determined upper limit the re-circulating valve will be automatically opened.

Condensers

The primary functions of a condensing plant for a turbine are:

1.   Extraction of air, non-condensible gases and vapours in the steam.

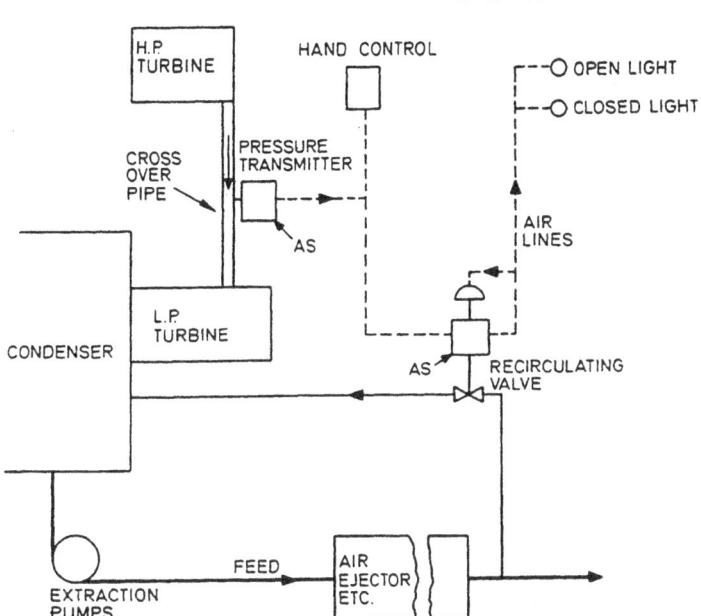

Fig. 6.10 FEED RE-CIRCULATION SYSTEM (PRESSURE)

2. To create the maximum vacuum by condensation so utilising maximum expansive steam work in the turbine.

3. To fully condense the steam with no undercooling of the condensate.

Theoretical and design aspects:

Dalton's Law

As applied for mixtures of air and water vapour.

1. The pressure of the mixture of a gas and a vapour is equal to the sum of the pressures which each would exert if it occupied the same space alone.

2. The pressure (and quantity) of vapour required to saturate a given space is the same for a given temperature irrespective of whether or not any other gaseous substance is present.

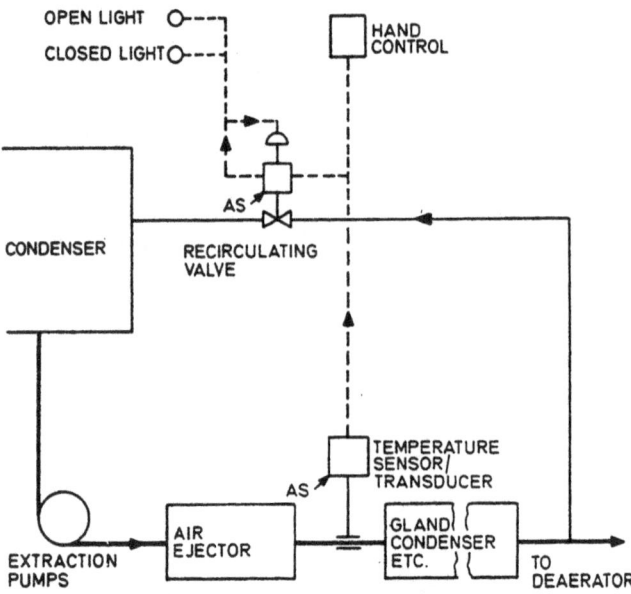

Fig. 6.11 FEED RE-CIRCULATION SYSTEM (TEMPERATURE)

Individual pressures in a mixture are known as partial pressures. Normally the steam entering the condenser will contain about 0.05% mass of air and the partial pressure of the air is very small.

Work done by steam

The lower the exhaust pressure (for a given admission pressure) the more the work done. Steam at 0.1 bar abs occupies approximately 14 000 times the volume it occupied as water so that condensation in a closed space creates a partial vacuum which can be much increased by extracting air and non-condensable gases which have leaked into the system.

Taking standard atmospheric pressure as 1.01325 bar

Condenser Efficiency =

(1.01325 − condenser pressure)/(1.01325 − condenser pressure corrected to circulating water outlet temperature) × 100

Vacuum Efficiency =

(1.01325 − condenser pressure)/(1.01325 − condenser pressure corrected to condensate temperature) × 100

Consider the following example for a modern condenser of the regenerative type:

Circulating water inlet temp.	21° C
Circulating water outlet temp.	28° C
Condenser pressure	0.05 bar
Condensate temperature	31.5° C
Condenser pressure corrected to 28° C	0.038 bar
Condenser pressure corrected to 31.5° C	0.045 bar

(Condenser Efficiency) =
0.96325/0.97525 × 100 = 98.8%
(Vacuum Efficiency) =
0.96325/0.96825 × 100 = 99.4%

Cooling surface required

Important factors are:

1.   Velocity of cooling water in tubes, this is mainly governed by the possibility of impingement attack and hence material choice. The velocities used give streamline flow of cooling water in the tubes, whilst reducing erosion attack this gives poor heat transfer compared to turbulent flow. By using titanium tubes water speed could be increased to give turbulent flow without impingement attack, thoughts are now being directed towards the use of a titanium type plate heat exchanger surface.

2.   Number of passes for the cooling water. A two pass using a circulation pump would give a water speed of about 1.85 m/s. Single pass with scoop circulation, water speed about 1.4 m/s—this depends upon vessel speed, head of water available and resistance to flow. Scoop intake means increased cooling surface required and a slight reduction in vessel speed.

3.   Cooling water temperature.

4.   Condenser tube material. Materials which can withstand impingement attack are hard and strong—they can be thinner but have poorer heat transfer coefficients and are expensive.

5.   Tube dimensions and distribution. Obviously large diameter tubes would give less resistance to water flow, reduced possibility of

fouling but smaller cooling surface area per unit volume. A typical size could be 19 mm outside diameter 18 wire gauge.

Distribution arranges the steam flow and air flow pattern and is affected by condenser intake *e.g.* from above, single axial entry, double axial entry. In addition, air extraction point, size and type of condenser well (affecting steam lanes for regeneration) are other points that must be considered.

6.   Cleanliness of tubes. With modern plant no steam space fouling should occur because of purity of steam and condensate. Water space fouling can and does occur hence a 'fouling allowance' may be made in cooling surface calculations.

7.   Quantity of air in the condenser and the capacity and efficiency of the air ejectors.

Steam consumption is about 3.7 kg/kWh and average air leakage is about 0.05% of this. *i.e.* 0.00185 kg/kWh.

This would mean about one air tube for every 2000 steam tubes for equivalent temperature drop. In a two pass condenser, air tubes are generally only single pass and the ratio of air to steam tubes would be 1 : 1000. The above assumes similar heat transfer rates. In practice air is a very poor conductor of heat, the overall heat transfer coefficient for condensing steam is about 5680 J/m²s° C and for air about 45 J/m²s °C. This means in practice the ratio of steam to air tubes is 45/5680 × 1000 : 1 *i.e.* 8 : 1 approximately.

Note how air is a very undesirable agent in a condenser from the heat transfer view point.

8.   The quantity of steam condensed per unit of surface.

The Grasshof formulae for log mean temperature difference (θ_m) is often used:

$$\theta_m = \frac{\theta_o - \theta_i}{\ln\left(\dfrac{\theta_s - \theta_i}{\theta_s - \theta_o}\right)} \,^\circ C$$

where θ_s, θ_o, θ_i refer respectively to steam, cooling water outlet and cooling water inlet temperatures, the hyperbolic law is followed (ln is hyperbolic log).

The area in m² of cooling surface (A) can be then evaluated from:

$$A = m.h_{fg} \, / \theta_m U$$

where *m* is steam condensed per hour, h_{fg} is latent heat in J/kg, *U* overall heat transfer coefficient J/m^2h° C.

Position of condenser and drainage

Underslung, or hung, type of one or two pass regenerative condensers are integral in some way with the l.p. turbine above them. They may be supported from above by a combination of beams, which support the l.p. turbine and spring type chocks from below. In the case of condensers supported on spring type chocks the weight must be fully taken up on rigid supports before water testing the condenser.

Alternatively, they may be completely hung from support beams which also support the weight of the l.p. turbine. Fig. 6.12 shows diagrammatically the arrangements. Drainage into the well at the bottom of this type of arrangement is little problem.

Axial flow types, either single or double pass, for single plane turbine installations (see Fig. 6.12) generally use a four point support system which allows for expansion (with location by keys) so as not to affect the l.p. turbine. Drainage with this type of plant is more of a problem (it would be disastrous if due to vessel movement condensate came back into the l.p. turbine due *e.g.* to extraction pump failure).

Alarm and emergency trips for condenser water level are provided and these can be easily tested by flooding the float chambers thus simulating the fault.

In some modern condensing plants a dry bottom condenser is used. This means that no water well reservoir exists in the condenser bottom. The actual water storage point in the system where water can be held and controlled could, for example, be the de-aerator quite easily, if required. The dry bottom condenser will have little undercooling, as the instant the steam is condensed and reaches the extraction pump it is removed. The extraction pump must be capable of 'running dry' for periods and 'flooded' for other periods. The pump will be briefly described later. Another innovation that has been tried for extraction of gland steam is to utilise electric fans. No real conclusions as to advantages or disadvantages of such an extraction arrangement can be evaluated at this time.

Construction

Modern condenser shell and water box ends are made of prefabricated steel plate. The water box ends have (after pickling, shot blasting and thorough cleaning) rubber or neoprene bonded to them. In some installations cast iron water box ends, internally coated with a protective compound will be found, but for modern highly rated plant with very large condensers such a casting would be difficult and expensive to produce.

Tube plates are generally rolled naval brass. Tubes solid drawn aluminium brass or cupro-nickel (usually the former). Baffles and support diaphragms steel plate.

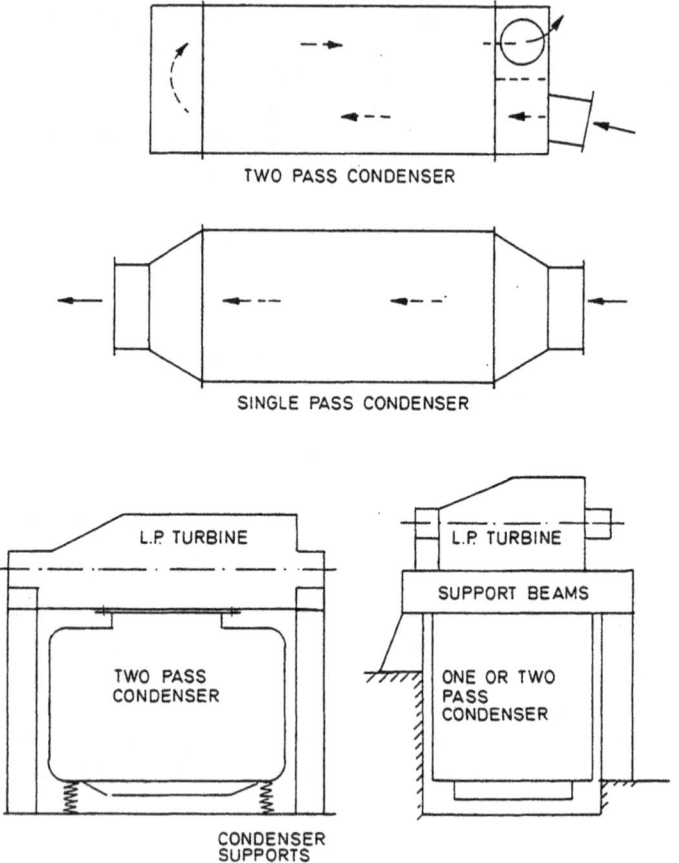

TWO PASS CONDENSER

SINGLE PASS CONDENSER

L.P. TURBINE

TWO PASS CONDENSER

L.P. TURBINE

SUPPORT BEAMS

ONE OR TWO PASS CONDENSER

CONDENSER SUPPORTS

Fig. 6.12

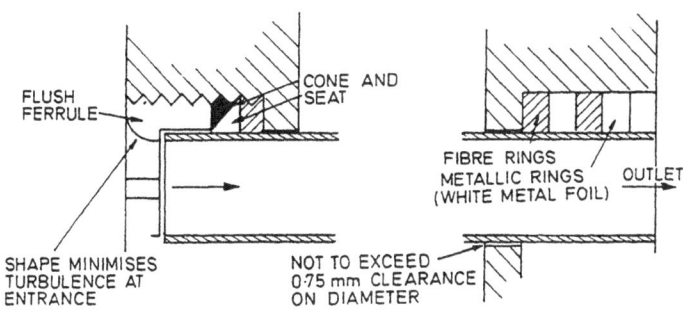

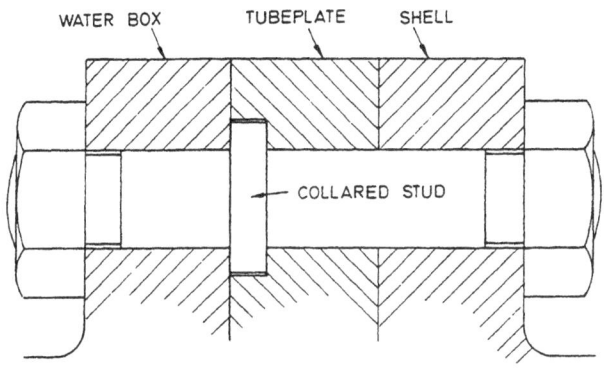

Fig. 6.13 CONDENSER FITTING DETAILS

Figs. 6.13 and 6.14 show some condenser fitting details.

Tube fittings: various methods are in use they are:

1.   Ferruling at the inlet end using a combination of fibre and metallic packing rings around the tube at outlet. This method is being superceded by one or other of the following the aim being to dispense with ferrules and economise.

2.   Expanding at inlet end forming a bell mouth to minimise impingement attack by providing a smooth entry, and packing rings around the tube at outlet. This arrangement allows for free expansion of the tubes relative to the plates.

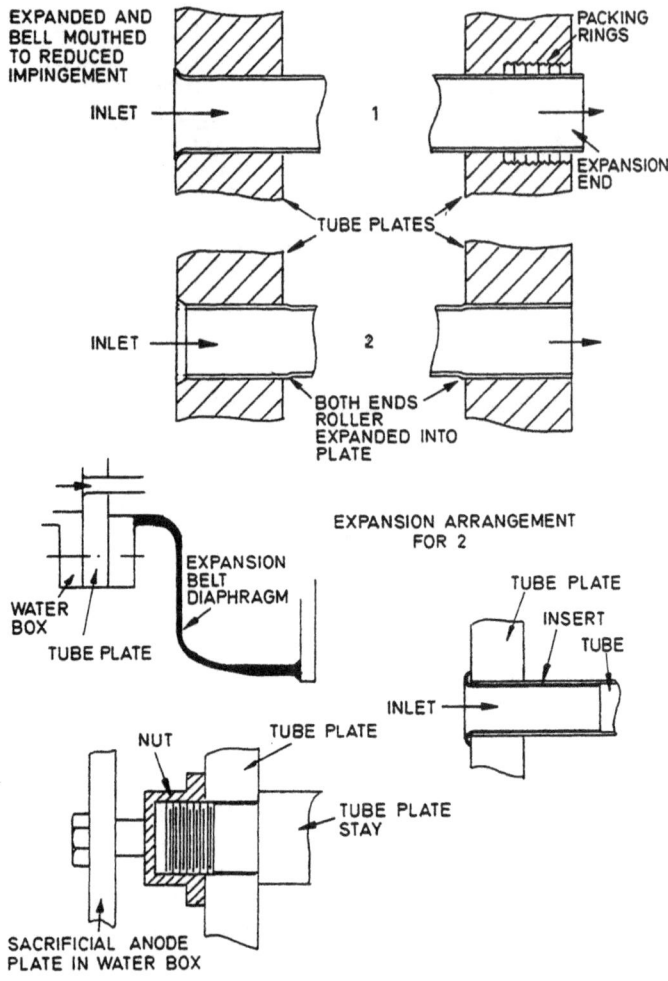

Fig. 6.14

3. Both ends of the tubes roller expanded into the tube plates—no packing or ferrules but allowance must still be made for expansion. This is achieved by using a expansion piece insert in the shell.

Regenerative Condenser
See Fig. 6.15.

The construction is such that a direct passage is left down the

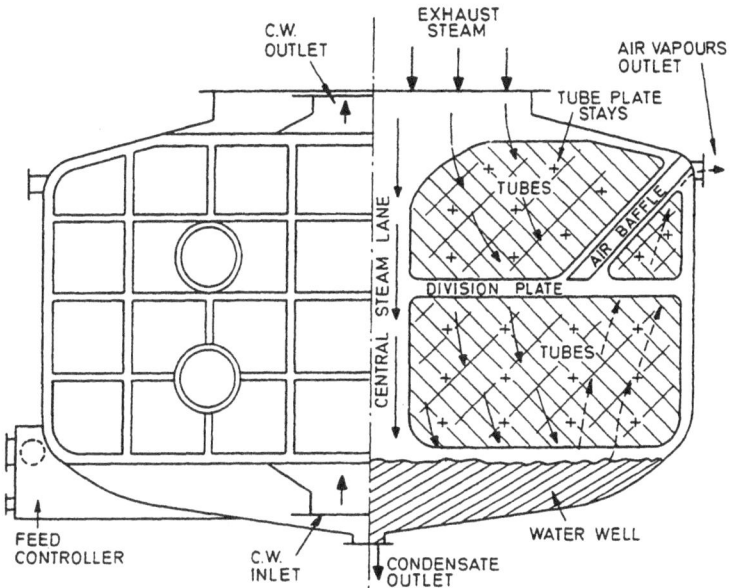

Fig. 6.15 REGENERATIVE CONDENSER

centre of the condenser. Exhaust steam can pass directly on to the feed water lying condensed in a large well at the condenser bottom. The level of water in the well being maintained and controlled by the closed feed controller.

The central lane allows direct contact between l.p. steam and condensate so giving regenerative heating. The condensate is then maintained at the same temperature as the exhaust steam, this means negligible undercooling and only latent heat extraction during condensation which is the ideal thermal principle.

Air vapours are driven ahead of the steam and subject to maximum cooling surface before being drawn up and extracted under the air baffles to the air ejector with minimum volume.

Undercooling of the condensate is of importance not only from the thermodynamic aspect but also gas absorption. If the sea water temperature falls, undercooling and gas absorption will increase with consequent loss in efficiency and increased risk of corrosion. Hence in modern practice control over cooling water flow either by quantity or re-circulation so that the condenser operates at design conditions at all times, is used.

The regenerative condenser underhung from the l.p. turbine is relatively common.

The neck of the condenser must be rigidly stiffened, free expansion occurs at condenser bottom on to spring chocks on the tank tops. A perforated baffle is often provided at the top to prevent direct steam impingement on to the top rows of tubes.

Defects, precautions, remedies

1. Corrosion

This generally takes the form of erosion/corrosion *i.e.* impingement attack. The large mass of non-ferrous metal (tubes and tube plates) would be cathodic in sea water to the steel water boxes and would therefore be protected by a coating of iron salt at the expense of the water boxes. However the water box ends are protected by rubber or neoprene and this means alternative protection must be given.

Sacrificial anodes of steel plate (Fig. 6.14) with or without impressed current may be provided in conjunction with ferrous sulphate dosage.

Ferrous sulphate dosage would normally be carried out, in an established plant, prior to entering port and upon leaving. River and estuarine waters generally contain abrasives which erode away the protective iron salts on the tubes and may in addition contain corrosive pollutants (sulphur compounds etc.). Once at sea, erosive attack is reduced and the sacrificial anodes should, after the initial ferrous sulphate dosage, give sufficient protection.

Ferrous sulphate can gradually build up in the condenser, impairing heat transfer as it does so.

To remove possibility of impingement attack at tube entry inserts made of nylon or P.T.F.E. can be used. The tube entry is first cleaned and degreased then the adhesive coated insert is pushed into place Fig. 6.14 shows briefly the arrangement.

2. Cracking of tubes

This could be caused by stress corrosion, tubes would have to have some stresses locked up in them due *e.g.* to bad manufacture.

A galvanic action takes place between regions of different stress leading to cracks and stress relief. Correct manufacture—including shop test using mercurous nitrate—and installation should avoid this defect.

3. Blockage due to marine growths

This can be prevented by using a chlorination system, or if a impressed current system is being used a biocidal sacrificial anode can be used in addition to the steel anodes.

Vibration

Hull transmitted vibration to the condenser and tubes could cause the tubes to resonate with possible necking at the fixed ends, necking and thinning in way of the division support plates.

It is not good practice to allow a tube length greater than 100 diameters between supports, also a support should not divide the tube length in an exact ratio so as to make the support a vibration node.

Leakage detection

Various methods are used if the leakage is not obvious.

1. Fluorescein Leak Determination

When a condenser is known to have leaky tubes leak detection using fluorescent dye simplifies the task. The condenser shell is drained and water containing the dye is slowly pumped in. This water will pass through any perforation in a tube wall or at the tubeplate, and emerge from the interior of the tube at the tube end. Ultra violet light illuminating the outer end of the condenser will cause the water to fluoresce and so allow easy detection of the faulty tube.

2. A ultra sonic generator placed inside the condenser 'floods' it with ultra-sound. By using a head set and probe, tube leakage can be homed in on. Where a pinhole exists sound 'leaks' through and where a tube is thinned it vibrates like a diaphragm transmitting the sound through the tube wall.

3. With slight steam pressure on the condenser, leakage of steam molecules through a pinhole in a tube generates ultra-sound which can be picked up by a probe and head set.

4. Plastic sheets placed over the tube plates will be drawn into a leaking tube with the condenser under vacuum.

General causes of loss of vacuum
1. Restriction of cooling water flow—choked filters, damaged circulating pump etc.
2. Leaking condenser water box division plate causing by-pass in a two pass condenser. Could be corrosion damage.
3. Rising condensate level in condenser—extraction pump failure, closed feed controller sticking etc.
4. Air leakage in glands, drains etc.
5. Defective air ejector.

Condensing plant performance

| Reading No. | Cir. Water | | | Condenser | | | REMARKS |
	Inlet temp. °C	Outlet temp. °C	Rise °C	Abs. Press. in bar	Steam Inlet °C	Condensate Temp. °C	
1	16.7	22.7	6	0.04	29	28.4	Very good performance
2	16.7	22.7	6	0.053	34	32	Dirty tubes
3	16.7	22.7	6	0.053	34	22.4	Defective air ejectors or air leak
4	16.7	22.7	6	0.067	38	28.9	Defective air ejector and dirty tubes
5	16.7	30	13.3	0.048	32	30	Insufficient Circulating Water

Note:
The effect of dirty tubes is a loss of vacuum and a rise in condensate temperature, whilst the effect of an air leak is a loss of vacuum and virtually constant condensate temperature. This can be established theoretically by applying Dalton's laws.

Air ejector

Consider the air ejector of the three stage type as sketched in Fig. 6.16:

The average feed temperature rise is about 5° C through the ejector. The vacuum at the nozzle, for each stage, is about 0.034 bar, 0.1 bar and 0.17 bar for a normal rate of air leakage (100 kg/h) and

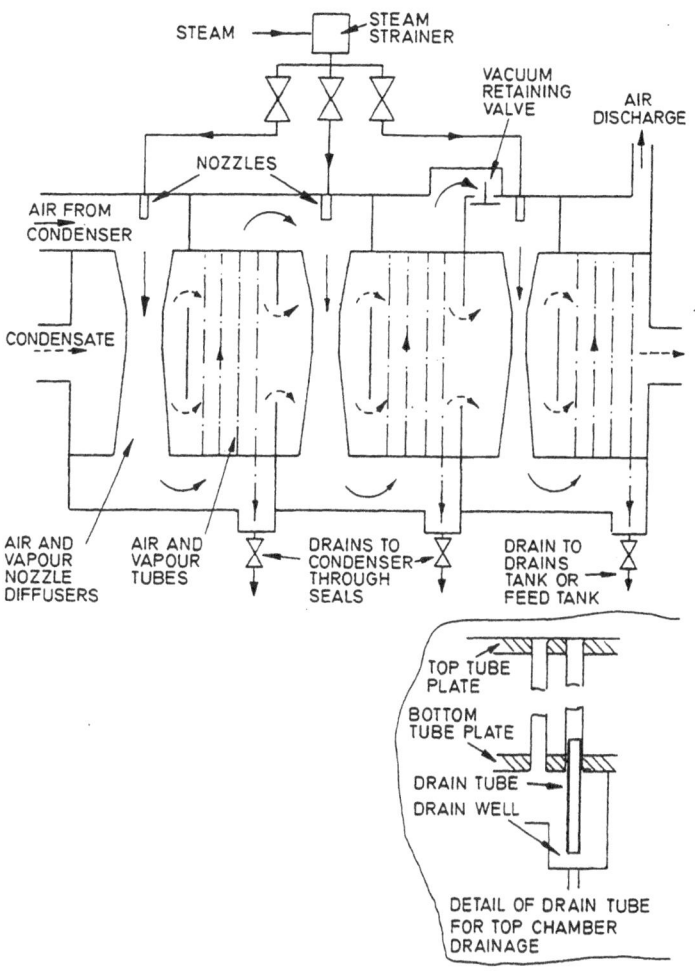

Fig. 6.16 THREE STAGE EJECTOR

760 mm Hg barometer reading. The steam jet velocity is about 1200 m/s and steam consumption about 0.5% of main engine steam consumption. The circulation path of feed water and steam and air vapours can be seen on the sketch. Note the following two points which are not detailed closely on the sketch:

1. The main vapour nozzles decrease in area from first stage inlet to third stage outlet, this is because the vapours have been cooled and reduced in volume.

2. The steam nozzles usually differ in area, in general the nozzles increase in area from first stage to third stage. The function of the first steam nozzles is primarily to produce a high velocity steam jet to obtain maximum vacuum whilst the function of the last nozzle is primarily to increase momentum so as to 'pump' out the contents against atmospheric pressure. This requires different nozzle design for each stage regarding velocity and mass flow.

The position of the vacuum retaining valve in relation to the drains is important. This valve must prevent any air return to condenser in the event of failure of steam nozzles. Any connections to the condenser on the condenser side of the ejector (i.e. to the left of the sketch) must be protected from air return flow. Hence the vacuum retaining valve must be placed after the second stage drain, the third stage drain need not be so protected if it leads to an atmospheric drain tank. If all three drains were led to the condenser the vacuum retaining valve would have to be at the foot of the air discharge pipe. To facilitate continuous drain removal (without direct connection between points at different pressures) the drain pipes are usually provided with U shaped seal pipes or float traps. The object of such seals is to prevent direct connection between two points at different pressures. For example at the second stage of the ejector the differential pressure across the drain may be 0.1–0.034 bar, i.e. 0.066 bar assuming condenser vacuum is 0.034 bar. This would require a loop seal with water of at least $0.066 \times 10^4 = 660$ mm H_2O head difference in the U loop to prevent direct connection.

Two stage air ejector

High velocity steam issuing from the first stage nozzle entrains air and vapour from the main condenser. Condensate circulating through the water box and alumininium brass 'U' tubes condenses most of the vapour and operating steam which is drained via a loop seal back to

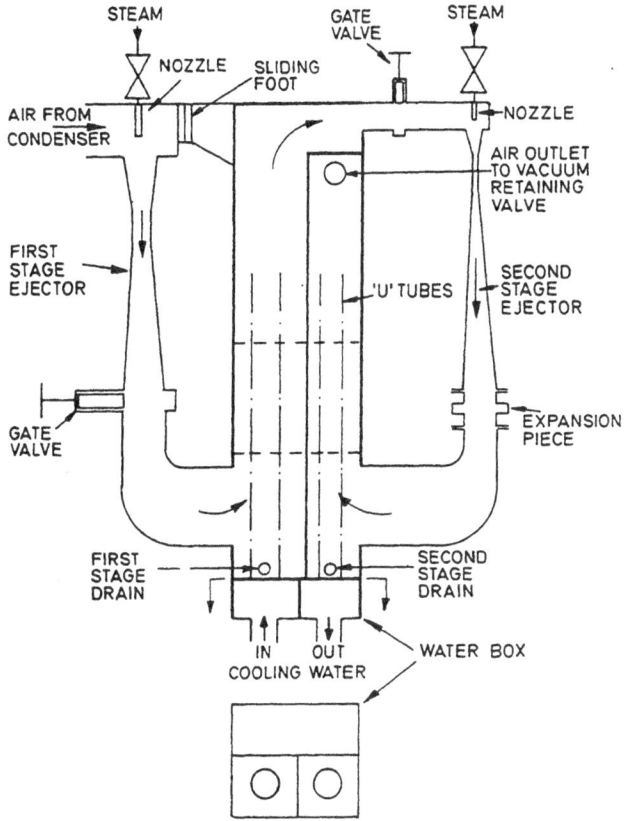

Fig. 6.17 TWO STAGE AIR EJECTOR

the main condenser. The air and remaining vapour pass into the second stage which discharges into the second stage ejector condenser wherein the remaining vapour and operating steam is condensed and drains back to the atmospheric drains tank. The air is discharged from the second stage through the vacuum retaining valve to atmosphere.

The unit is generally provided with two first and two second stage ejectors but only one pair would be in operation. Relief valves are fitted to both stages and the gate valves have a gland water seal which gives indication of gland condition.

Pumps

The pumps in the system are either electric or steam turbine driven, either vertical or horizontal, centrifugal type. Basic principles of such pumps have been covered in Vol. 8. From a safety aspect condensate pumps must be interconnected and they are usually arranged so that failure of one pump will cause an alarm to blow and the other pump will be auto-started by means of a pressure switch. Similar remarks apply to feed pumps.

A short description of a condensate extraction pump and a turbo feed pump will now be given.

Condensate extraction pump

A modern design is capable of running with suction water at a temperature equivalent to the boiling point at the pressure existing in the condenser. A typical unit runs at 175 rev/min, is vertically mounted, with two impellers discharging through diffusers for greater efficiency. The pump is arranged to be self regulating by the free suction head controlling the rate at which water is pumped. If the quantity of water reaching the suction is decreased this will reduce the pipe suction head and reduce the discharge pressure. Discharge quantity then reduces until it equals suction quantity. At the designed

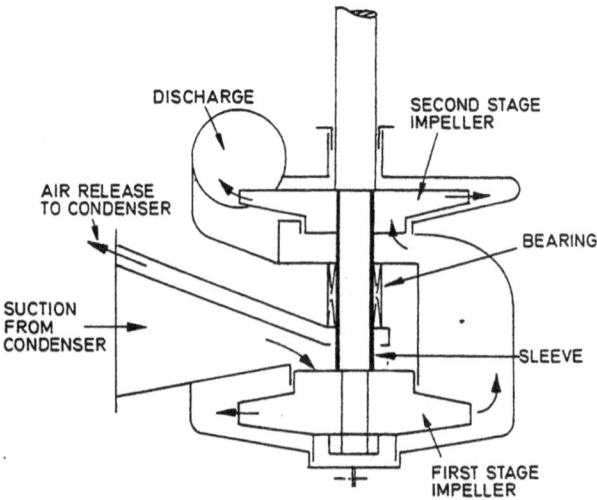

Fig. 6.18 TWO STAGE EXTRACTION PUMP

minimum suction head the pump will stop discharging. It is unnecessary for a condenser water level control device and the pump can be worked with dry bottomed condensers. The first stage impeller handles suction water near boiling point at high vacuum and discharges to the second stage impeller suction at above atmospheric pressure. Any vapour formed in the eye of the first stage impeller is vented back to the condenser.

Turbo feed pump

A diagrammatic sketch of the pump including control devices is shown in Fig. 6.19.

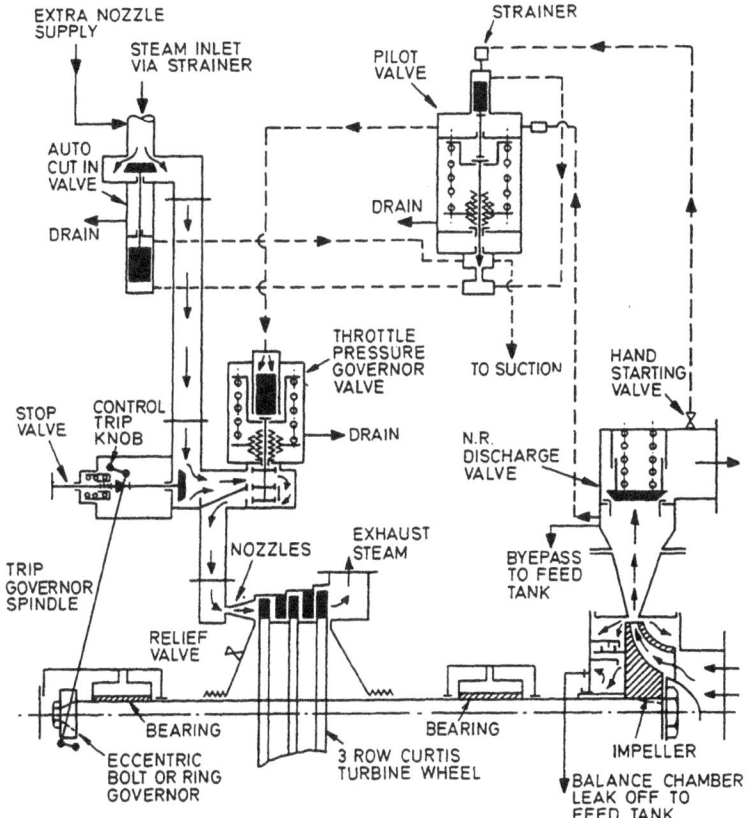

Fig. 6.19 TURBO FEED PUMP SYSTEM

For *examination* descriptive purposes the sketch could be simplified by omitting the detail of all five valves and merely sketching them as diagrammatic rectangular figures.

Referring to the sketch:

The pump in this stage has a single stage impeller (monel metal) driven by a turbine wheel having one pressure stage and three velocity stages. The shaft (chrome nickel steel) is supported on white metal lined journal bearings lubricated by oil rings and fresh or salt water cooled. Turbine glands are of the carbon ring type and may have gland sealing supply steam if exhaust is to a vacuum. Most of these pumps are designed to work with a positive suction head, a minimum of say 1.2 m for self-priming, for example about 7.5 m head of water supply is provided for water at 104° C. The pump when operating has a steam thrust effect axially but this force is small compared to hydraulic forces and is easily balanced. End forces are maintained in equilibrium by the balance chamber. Any movement to the suction side reduces the balance chamber supply gland clearances and increases the balance chamber leak off clearance so reducing balance chamber pressure. Movement to the turbine end increases balance chamber pressure until equilibrium again exists in an axial direction. Total axial clearance of the pump is correct at about 0.3 mm but this is much greater on the turbine blading end (5 mm total). The by-pass is provided on the discharge side of the pump to maintain a circulation at light load conditions.

The governor is usually of the eccentric ring type. When revolutions become excessive (500 rev/min above normal) the ring moves out under centrifugal force against the action of a spring, thus movement is transmitted by a trigger and levers so that the stop valve is unlocked and closed by the spring.

The throttle pressure governor valve has a *down* force due to pump discharge pressure (maximum at no load) which works against the *up* force of an adjustable spring. Equilibrium being attained under all load conditions.

The automatic cut in gear is provided where the pump is used for standby duty and functions to automatically start the pump when feed line pressure falls below a certain value. With the pump on standby the stop valve is open and pump in readiness for starting, but as discharge pressure cannot release through the pilot valve it acts on the *underside* of the piston on the auto cut-in valve and keeps the

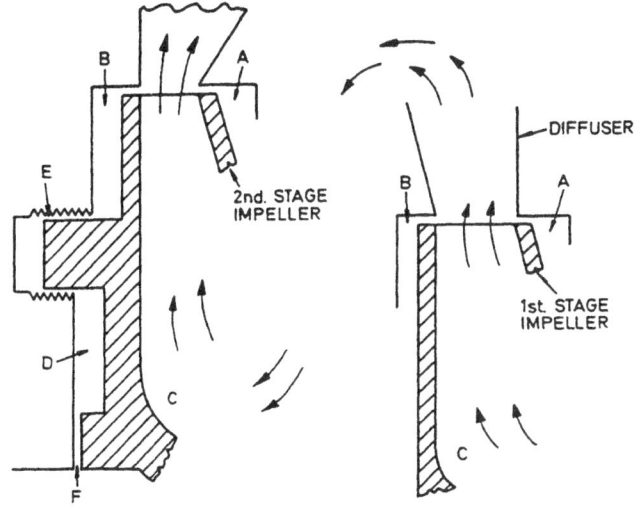

Fig. 6.20 TURBO FEED PUMP HYDRAULIC BALANCE

steam valve on its seat. If discharge falls below say 3.5 bar (but higher than boiler pressure) of the normal discharge pressure the spring of the pilot valve causes this valve to open. This releases pressure on the auto cut-in valve through the pilot valve to suction and the steam valve admits steam to the unit. Once the pump is running the other connection from under the discharge valve seat maintains the pilot valve open by acting on the underside of the top piston and the pump continues to run. This connection also leads to the throttle governor valve which functions in the normal way. The pump is stopped by closing its stop valve which in turn closes the pilot valve and auto cut-in valve, the stop valve can now be re-opened ready for emergency duty again. The hand start valve is provided to test the auto cut-in device or to enable the standby pump to be started by hand if required, for standby duty it is of course open.

Hydraulic Balance

The shaft assembly is maintained in its correct axial position by the hydraulic balancing arrangement. This arrangement is incorporated within the pump and is effected by pressure variations acting on the impellers taking into account the steam thrust on the turbine wheel.

In each stage of the pump areas A and B are subjected to the appropriate discharge pressure and areas C to their corresponding suction pressure. The total resultant thrust on the shaft due to these differences in pressure and area, tends to push the shaft towards the turbine end. This resultant thrust is opposed by an opposite thrust acting on the back of the second stage impeller, due to the pressure in D, plus the steam thrust on the turbine wheel. This pressure in the balance chamber D varies according to the rate of flow of the balance chamber leakage from B through the gaps E. Movement in either direction will continue until the opposing thrusts are equal; the pump is then said to be in hydraulic balance.

Variation in thrust will move the shaft, thereby changing the width of gap F. As F is increased or decreased, the pressure in D will correspondingly rise or drop until the hydraulic balance is again restored. This action is entirely automatic and practically instantaneous, maintaining the axial position of shaft assembly within 0.02 mm and enabling the pump to operate under all conditions without manual adjustment of the inner clearances.

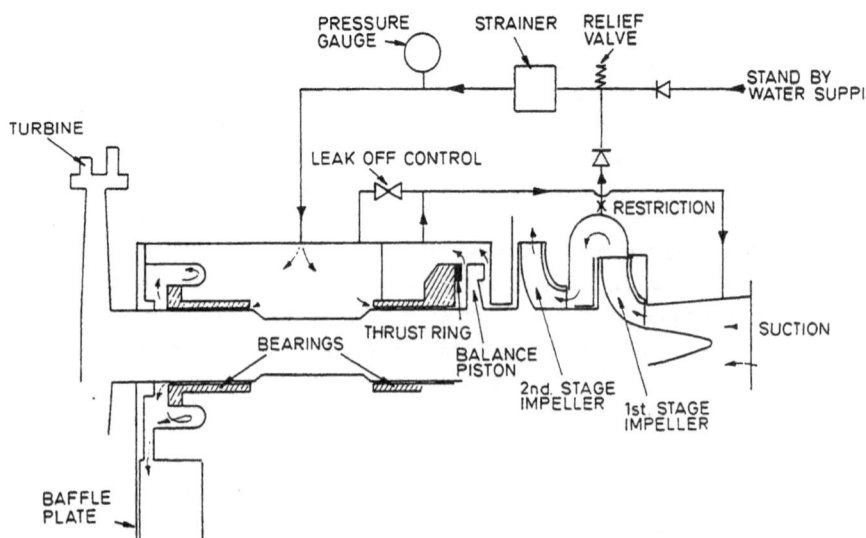

Fig. 6.21 WATER LUBRICATED FEED PUMP

When starting up, axial thrust of the steam flowing through the turbine pushes the shaft towards the pump end. A thrust ring at the pump end bearing absorbs this initial load and limits the travel of the shaft during the starting periods. Once the pump is running, the faces of the shaft collar and the thrust ring will not come into contact.

The balance chamber leakage combines with the lubricating water from the pump end bearing to be led via a pipe to the suction branch. When the pump is running there must always be a leakage through this pipe, otherwise severe damage to the unit will occur.

Turbine driven water lubricated feed pump

The casing for turbine and pump is one unit of cast steel with end covers for access to turbine and pump impellers, etc. The turbine consists of a simple two row, velocity compounded, high speed, overhung impulse unit secured to a stainless steel shaft. The pump at the opposite end of the shaft is made up of two highly polished stainless steel impellers. In way of the bearings the shaft is stepped and chrome plated, bearings are split bronze bushes P.T.F.E. impregnated in mild steel housings which are secured to the casing. Modern practice with P.T.F.E. bearing units in pumps, etc. is to leave the shaft journals slightly rough machined to that as they run in, transfer of some P.T.F.E. to the journals takes place, which reduces friction upon starting.

Bearing lubrication is by feed bled from the first stage impeller discharge through a multi-plate restriction device, non-return valve and strainer. If the water pressure becomes too high the relief valve diverts some to the drains system. As wear down takes place in the bearings an increasing quantity of water passes through them and hence pressure in the supply chamber will fall. A leak-off valve, fully open when bearings are new, can be closed in to reduce leak-off from supply chamber to pump suction and hence restore chamber pressure. Water leaking from the turbine end bearing combines with the exhaust steam.

For stand by duties a water supply from an external source such as condenser extraction pump, de-aerator storage tank, etc. would be required.

Control devices incorporated with the pump are:

1.   Overspeed trip in the form of a triggered valve which shuts off steam supply.

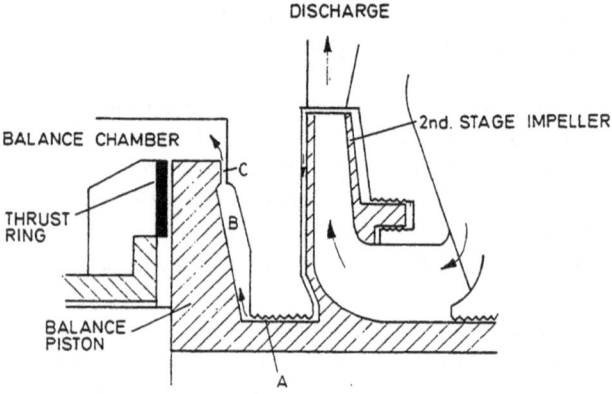

Fig. 6.22 HYDRAULIC BALANCE

2.   Differential pressure governor for pump speed control, this gives better economy than the discharge pressure type.

Hydraulic balance

The hydraulic balance arrangement maintains the rotor in its correct position, within limits, with minimum axial movement under all load conditions.

When starting the pump, the rotor may be forced towards the turbine end against the non-metallic thrust ring, but the turbine thrust (which is small) and the hydraulic thrust on the impeller move the rotor away from the turbine end. Water passes from the second stage through the restriction A into space B. The water at a reduced pressure acts on the balance piston forcing the rotor towards the turbine end. As the gap C widens the pressure in space B falls and the rotor moves towards the pump end until balance is restored.

Water pressure in the balance chamber is maintained at pump suction pressure by the leak-off connection, which allows water to flow back to pump suction. It must be appreciated that this pump has:

1.   No glands.
2.   No oil—hence no oil pump and cooler.
3.   Simplified hydraulic balance.
4.   Simplified construction.
5.   Improved control.

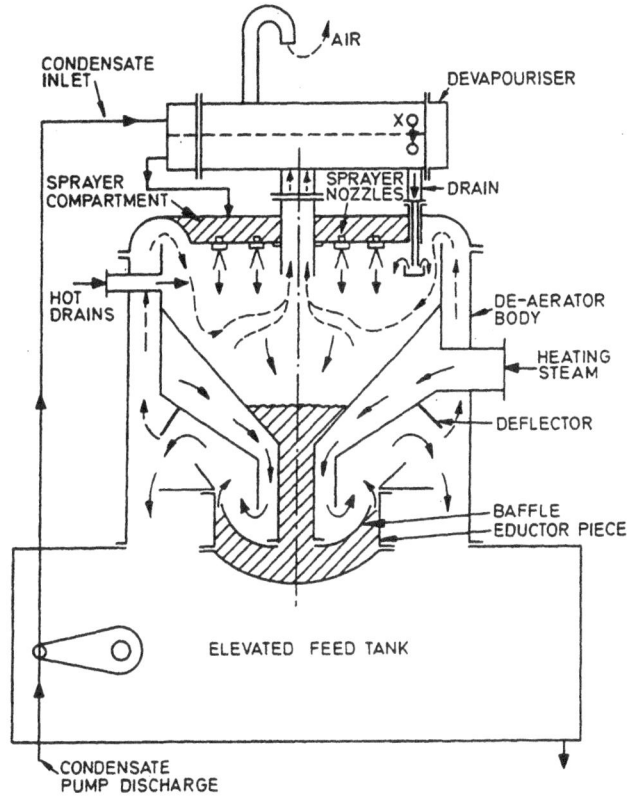

Fig. 6.23 DE-AERATOR

De-aerator

The operation of this unit is a direct development from the principle of the d.c. feed water heater described earlier.

Referring to the de-aerator sketch: Fig. 6.23.

Condensate enters the devapouriser, gaining heat from the rising air from the de-aerator body which passes over the outside of the tubes. The devapouriser is two pass and the pipe X serves to pass vapour and air to the upper compartment, air going to the atmosphere and further condensed vapour draining back through X.

From the devapouriser the condensate enters into the sprayer compartment. This condensate is then sprayed through nozzles into the main body of the de-aerator where it is heated by direct contact with hot drain exhaust. The sudden pressure drop and heat gain effectively liberates the air (compare this principle to that of the d.c. heater previously described). Condensate and exhaust steam finally mix and flow down to the elevated feed tank at a temperature approximately that equal to the saturation temperature of the steam (this *may* be below the temperature boiling point at atmospheric pressure if the exhaust steam is at a slight vacuum, but this depends on the steam exhaust system supply and de-aerator and vapouriser extraction system design).

The air flow, at separation, is upwards so getting progressively colder, giving up heat and reducing in volume.

The de-aerator is by no means a standard unit as many closed feed systems do not incorporate this auxiliary but as boiler pressures tend to rise the de-aerator becomes almost a requirement.

Fig. 6.24 shows a modern de-aerator-heater which can operate at a relatively high pressure, *e.g.* 3.5 bar, and hence does not have to be situated in an elevated position.

The feed water is sprayed out of the non-ferrous nozzles in the de-aerator head in the form of a hollow atomised cone which presents a large surface area to the heating steam and is thus rapidly raised in temperature. These two things, the atomisation and heating, liberate very rapidly oxygen and other gases from the feed.

A collecting tray formed by the upper part of the steam belt allows the water time to boil, thus driving out the remaining gases before the water passes through the downcomers into the storage tank. The de-aerator head can be mounted on to a vertical or horizontally arranged storage tank, which has internal baffles to minimise rolling effects, the two forming a central piece in a packaged feed system.

Temperature rise of the feed water is at least 28° C with the oxygen content being reduced from 0.2 ml/l to 0.005 ml/l.

Other auxiliaries

None of the other auxiliaries in the system requires particular consideration further to that given. Most units are heat exchangers, some with air extraction devices, the principles of which have been covered previously.

Relating to heat exchangers one aspect could be simply con-

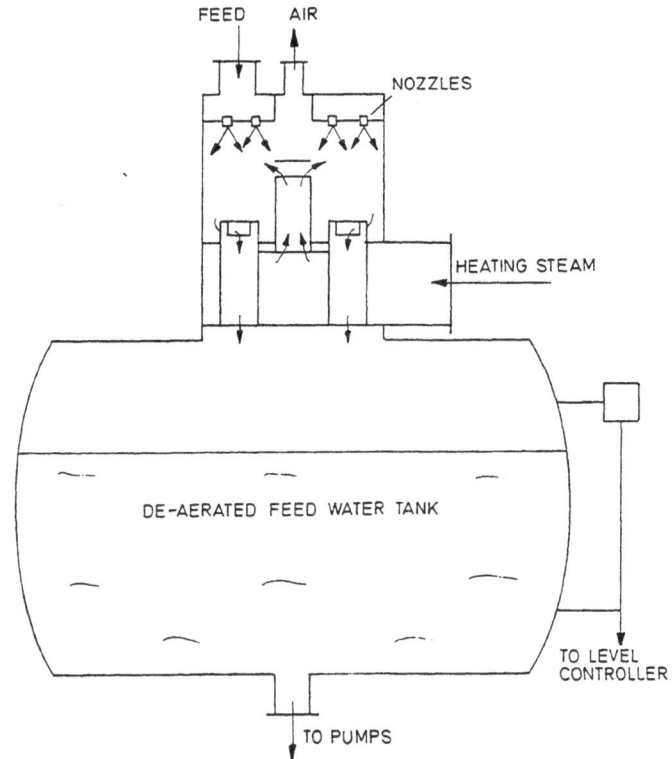

Fig. 6.24 DE-AERATOR WITH STORAGE TANK

sidered in conclusion namely that of the simple heat balance.

In the case of a feed water heater: Each kg of steam condensing from dry saturated steam with little undercooling would give up about 2100 kJ of heat (approx. latent heat) to the feed water. Taking the feed temperature rise at about 50° C then

Heat lost by steam = Heat gained by feed water
$$1 \times 2100 = m \times 4.2 \times 50$$
taking the specific heat of water as 4.2 kJ/kg° K

Hence the mass of feed water (m) would be 10 kg for each kg of steam. This means 0.1 kg of bled or exhaust steam is required per kg of water.

In the case of an economiser:

Heat lost by gas = Heat gained by feed water
$$1\tfrac{1}{2} \times 200 = 1 \times 4.2 \times \theta$$

Where the specific heat of gas is about 1 kJ/kg° K, gas temperature fall 200° C, water temperature rise $\theta = 80°$ C, approximately $1\tfrac{1}{2}$ kg gas in circulation per kg water.

In the case of an air heater:

Heat lost by gas = Heat gained by air

Where air and gas quantities and specific heats are approximately the same, temperature changes are approximately the same.

RELATED EQUIPMENT

Evaporators

The basic information given on evaporators in Vol. 8 should be considered. As the evaporator is perhaps a more important part of a steamship plant some further details are now considered. However, questions relating to same mainly occur in the E.K. general exam.

Single effect evaporation

Refer to Fig. 6.25 which is a slightly more detailed circuit than that previously considered. The flow rates given and temperatures given are typical for such a plant giving about 40 tonnes per day.

This particular plant gives a very good performance ratio utilising waste exhaust at about 0.6 bar for coil steam. It also has the big advantage of extremely small scale formation due to the high vacuum maintained in the shell.

Multiple effect evaporation

With double effect evaporation two evaporators are used in series with shell pressures at about 0.6 and 0.17 bar. The vapour from the first effect is used as the heating medium in the second effect. The arrangement is similar to the single effect plant shown in Fig. 6.25. Such complex plants are being superceded by simpler more economic arrangements.

Triple effect evaporation has been utilised for outputs of about 300 tonnes per day at 1.35 bar, 0.6 bar and 0.17 bar shell pressures at a

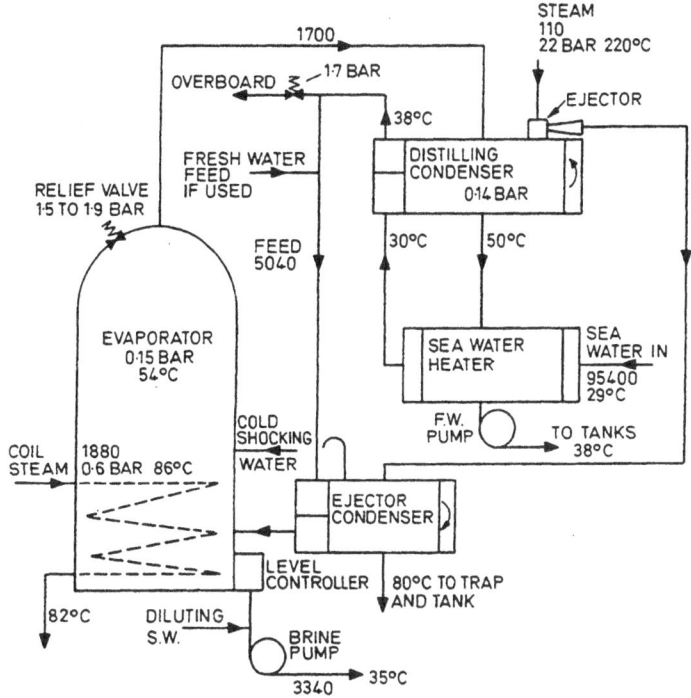

ALL QUANTITIES IN kg/h
FLOWMETERS AT ALL MAIN POINTS

Fig. 6.25 SINGLE EFFECT EVAPORATION

cost of about 1.1 tonnes of fuel oil. In this case a booster pump would probably be fitted after the dosing tank to force feed into the 1st stage evaporator.

Flash evaporation

This principle has been utilised in R.N. practice and is becoming increasingly used for M.N. vessels. The principle should be seen by referring to the sketch. The essential fact is that hot water has its pressure dropped quickly so that some water flashes to steam vapour.

The unit has the following advantages: Low brine density, boiling water not adjacent to heating surfaces, low scaling rate, self

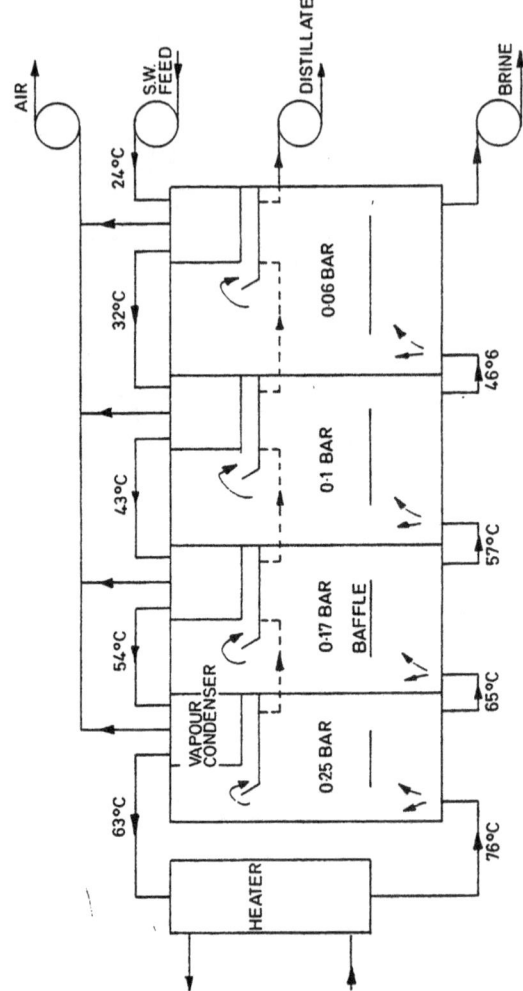

Fig. 6.26 FLASH EVAPORATION PLANT

regulating, efficiency increases with number of stages, compact. Note that the vacuum chambers increase progressively in size due to the increasing vacuum.

Feed treatment

   For continuous running of evaporators feed treatment may be

utilised. Chemicals must be automatically metered into the feed supply at a dosing tank, or pre-mixed with fresh water before addition to the feed.

Typical chemicals used have been: Chlorides or sulphates (granulated) of zinc, aluminium or ferric. Ferric chloride has shown good results in reducing magnesium hydroxide scale (occurring above 80° C saturation temperature in the shell) and calcium carbonate scale. Some trouble has been experienced with attack to piping and tanks and the latest practise is to utilise polythene linings. Refer to Vol. 8 for modern treatment.

TEST EXAMPLES

Class 2

1. Describe a closed feed system, detail the pressures and temperatures at the main points. State the advantages of such a system. Describe the functioning of control devices in the system during manoeuvring conditions.

2. Describe a turbo feed pump. Enumerate, with details, the automatic safety devices incorporated and state their particular function.

3. Sketch and describe a three stage air ejector as fitted in the closed feed system. What is the function of the ejector? What is the approximate rise in temperature of the feed water in passing through this unit? State the vacuum at each stage and indicate clearly on your sketch the position of the vacuum retaining valve and the positional arrangement of the drains.

4. In the case of a turbine driven rotary feed pump explain and show by diagrams how (a) the output of the pump is automatically governed, (b) the axial thrust is taken up when running normally, and (c) the axial thrust is taken up on starting.

5. Sketch and describe a surface feed water heater. Show clearly on the sketch the position of all mountings. State the advantages of feed heating.

Class 1

1. Sketch and describe a multi-stage evaporation system stating any advantages claimed for such a system compared to single stage. Clearly indicate on your sketch the pressures and temperatures at the

various points and give a reasoned explanation of the advantages of low pressure operation.

2. Sketch and describe a contact feed heater and name the connections and mountings. State the source of the heating steam. Where is the heater placed and why? What attention does the heater require to maintain it in an efficient working condition?

3. Sketch and describe a three stage air ejector. Detail the pressures at each stage. What maintenance is normally required? Do vapour nozzles and steam nozzles differ in size from stage to stage? Comment on this.

4. With the aid of sketches show how a closed feed system condenser water level is maintained under varying conditions of load. With the aid of a single line sketch describe a closed feed system flow of condensate with particular emphasis on the feed heating arrangement.

5. Describe a de-aerator. Detail the working pressures and temperatures. Where is the unit placed in the feed system and what arrangements are adopted to utilise the de-aerator for stand-by harbour use?

6. Discuss, in detail, the application of regenerative feed heating. Explain clearly how this improves plant efficiency. Make a simple line sketch of any type of feed water heater.

7. Describe, with the aid of sketches, a turbine or motor driven rotary feed pump suitable for working in conjunction with high pressure boilers. Explain clearly the operation of starting the pump and putting it on load.

8. What is meant by flash evaporation? Describe the theoretical principles involved. Make a line diagram of a suitable system inserting the pressures and temperatures at each stage.

MISCELLANEOUS SPECIMEN QUESTIONS

CLASS 1

1. Sketch and describe a three element feed regulator. Explain how unity relationship is maintained between the identified variables. State why three element control is superior to two element control.

2. With reference to main boilers explain:

(a) why membrane walls are less susceptible to distortion than separate fire row tubes,

(b) how reheat elements are protected against overheating under manoeuvring conditions,

(c) why headers have largely superceded water and superheater drums and define their functional value.

3. With reference to main superheaters:

(a) compare the advantages and disadvantages of contra-flow with parallel flow designs,

(b) describe with sketches how the element banks are supported,

(c) describe how boiler priming effects superheater effectiveness and condition.

4. Draw in detail a section through a consolidated high lift safety valve, labelling all the principal components. Explain how the action of this valve differs from other types of spring loaded valve. Explain why this design was developed and to what extent the objectives have been achieved.

5. With reference to main boilers compare, using sketches, the design details and merits of:

(a) stud and membrane furnace walls,

(b) water and air cooled desuperheaters,

(c) gas and steam air heaters.

6. Describe with sketches two ways whereby the steam temperature at superheater outlet is strictly controlled under all conditions of boiler operation. Discuss the merits and demerits of the two ways selected compared with the other means of achieving the same purpose.

7.   Describe with sketches a reheat boiler incorporating a steam cooler, an economiser and air heater. Explain the purpose of the reheat cycle and state what are its advantages and disadvantages compared to simple cycles. Explain the problems peculiar to reheat boilers.

8.   Make a diagrammatic sketch of a non-reheat radiant boiler incorporating two superheater banks, a steam cooler, an economiser and an air heater. Give your reasons for adopting the layout you have shown.

9.   Describe with sketches a triple/double reduction gear for a set of cross compound turbines. State why combinations of epicyclic and parallel shaft reductions are occasionally employed. Suggest why combinations of star and planetary epicyclic gears are also used.

10.   Explain the following terms when related to gear tooth condition:

(a)   polishing,
(b)   surface fatigue,
(c)   scuffing,
(d)   flaking.

11.   Explain what is meant by 'nodal drive' and the function it performs. Describe with sketches a main gear train in which this arrangement is used. Give two advantages and one disadvantage of this arrangement.

12.   With reference to the examination of main gearing:

(a)   explain how undercutting or pitting of the teeth may be measured so that further development of these troubles may be easily measured,
(b)   describe a method of recording tooth surface condition,
(c)   state why such records are necessary.

13.   Sketch and describe a main propulsion cross-compound turbine installation constructed on a single plane arrangement with axial exhaust.

14.   Explain how a lubricating oil system of the direct pressure type differs from the gravity type. State two advantages it has over the gravity type. Describe how the main turbines and gearing are protected against total pump failure in a direct pressure system.

15.   Describe, with sketches, a single cylinder main propulsion turbine for say 15 000 kW. State what limitations are imposed on power

output and rotor speed for a turbine of this design and what advantages it possesses within these limitations.

16. State what are the causes and consequences of vibration in turbines. Describe how turbine vibration is eliminated, particularly in main propulsion units.

17. Give one reason for the cause of each of the following problems relating to turbine blading:

(a) fouling by deposits.
(b) erosion,
(c) rubbing of moving blades,
(d) vibration.

State one way whereby each problem is countered.

18. Explain why a main high pressure turbine is generally of the impulse type. Explain what is meant by the term 'gashed' rotor. Give three advantages and one disadvantage of this type of rotor in comparison with any other type.

19. Describe with sketches the emergency steaming connexions for a set of cross compound turbines. Explain how main propulsion may be restored with the high pressure turbine isolated. State with reasons what precautions and operational adjustments are necessary under these conditions. State what would be the difference if conversely the low pressure turbine was isolated.

20. With reference to blades for low pressure turbines state:

(a) how they are sometimes protected against erosion,
(b) why the cause and effect of erosion differs from that in a high pressure turbine,
(c) why shrouding is sometimes omitted from the last few stages,
(d) what is the function of lacing wire and why it is invariably used although rarely used in high pressure turbines.

21. Identify those factors which promote failure in turbine blades. State with reasons which factors induce rapid failure and those which have a long term effect. Describe the measures taken before and during service to counter blade failure.

22. Sketch and describe a turbine rotor/gear pinion fine tooth coupling. Give the advantages claimed for it over the coarse tooth coupling.

23. Show by a line diagram how protection is afforded to turbines

against overspeed, excessive axial movement, loss of vacuum or lubricating oil pressure. Explain how the system functions under fault conditions. Describe how the system is kept in a full state of readiness at all times.

24. Explain with sketches how a high pressure turbine designed for a reheat cycle differs from the non-reheat design. Give reasons for the differences found in the reheat turbine. Give two advantages of the reheat turbine.

25. Draw in detail a group of main turbine nozzle valves incorporating servo motor operation. Describe how selective/sequential operation is carried out.

26. Draw a line diagram of a complete gland steam system associated with a set of cross compound turbines, labelling the principal components and indicating the direction of flow in all lines. State how water consumption is kept to a minimum through its design and operation. Differentiate between its mode of operation under 'full away' and manoeuvring conditions.

27. Describe the arrangement and operation of a mixed bed demineralisation plant using cation and anion resins making special reference to:

(a) the ion exchange process,
(b) regeneration of the mixed bed,
(c) precautions in handling strong acids and alkalies and in the disposal of effluent.

28. Describe an automatic sootblowing system giving with reasons the sequence of operation. Explain why boiler tubes and superheater elements should be kept clean externally. State with reasons what precautions are taken during soot blowing operations. Explain how badly maintained soot blowers can cause boiler defects and loss of efficiency.

29. Draw in detail a main air ejector showing the flow paths and drainage arrangements. In particular explain why:

(a) a non-return valve is usually fitted on the vapour side,
(b) the nozzles and tubes are differing in diameter.
(c) cooling is necessary,
(d) performance tends to 'fall off' and how this is countered.

30. Analyse three important factors limiting the expansion of steam in main turbines. Explain how main condenser condition affects plant

efficiency. Give three causes of unsatisfactory condenser operation and state how they are dealt with. Describe with sketches how differential expansion between shell and tubes is accommodated if the tubes are expanded into both tube plates.

31. Sketch and describe a pressure controller as fitted in a steam range where the ultimate state error and response time are minimal. Identify those components giving the required characteristics.

32 Explain why turbine rotors are both dynamically and statically balanced after construction or reblading. State how rotors become unbalanced in service. Describe how a vessel is enabled to complete its voyage under its own power if vibration develops in one of its turbines.

33. Describe in general terms the principal features of a closed feed system. Explain how feed flow is controlled with variation of condensate in the de-aerator. State what is meant by 'dead band' in the control range and its function.

34. State with reasons what are the shortcomings of the conventional intermittent sampling and analysis of boiler water. Describe with sketches an arrangement for overcoming these inadequacies.

35. Sketch and describe a rotary cup burner for a boiler. Explain why this type of burner is suitable for use with automatic combustion control. Explain the meaning of the term 'turn down' ratio and give a representative figure for this type of burner.

36. Draw a line diagram of a boiler combustion control system labelling the principal items. Explain how the system functions and in particular how feed water supply, fuel supply and air/fuel ratio are regulated to match steam pressure and flow variation. Explain how these controls can be tested for alarm conditions without upsetting the balance of the system.

37. Compare the effects of controlling main propulsion turbines by either throttling through the manoeuvring valve or selecting the best nozzle combination.

38. Explain why leakage of steam into a low pressure turbine casing does not reduce the vacuum whereas leakage of air will. Sketch and describe a gland steam system for a set of main propulsion turbines. Explain how the glands are sealed under both full power and manoeuvring conditions and how steam leakage to atmosphere is kept to a minimum.

39. Describe three ways, commonly adopted, to improve the basic overall efficiency of a main propulsion plant for a given set of steam conditions at superheater outlet.

40. Give two reasons why a main turbine rotor may bend. Explain why this fault constitutes a hazard. Prepare operating instructions as a safety precaution against the possibility of rotor distortion.

41. State what are the indications and possible consequences of both significant and insignificant steam leakage into astern turbines during ahead operation. Describe tests to measure the leakage from an astern manoeuvring and guard isolating valve. State why some astern turbines are equipped with a high temperature alarm and how it is tested.

MISCELLANEOUS SPECIMEN QUESTIONS

CLASS 2

1. With reference to combustion state the effects of:

(a) fuel temperature and air supply on flame condition and funnel emission,

(b) salt water contamination of fuel,

(c) molten ash in combustion gases.

2. With reference to boiler safety valves explain the purpose of:

(a) easing gear,

(b) valve drains,

(c) adjustable ring seats,

(d) high lift facility.

3. Draw in detail a boiler feed regulator controlled by variables additional to water level. Describe how it operates. Give reasons for these additional parameters in feed control.

4. Draw a line diagram of a main boiler labelling the principal items and indicating the path and direction of flow of the combustion air and gases, the boiler water and steam. State how water circulation can be reduced to zero or even reversed. Give three reasons why the combustion gases are constrained to follow the route indicated.

5. With reference to main boilers:

(a) describe with sketches two ways of controlling superheat temperature,

(b) give three reasons why one arrangement is preferable to the other.

6. With reference to main boiler furnaces explain with sketches the arrangement and the purpose of:

(a) roof firing,

(b) tangential firing.

Give with reasons two disadvantages of both (a) and (b). Explain how (a) and (b) have affected boiler design.

7. Give two reasons for the regular systematic inspection of the

main gearing. Describe three operational faults to which main gearing is prone. State the probable causes and the methods by which they might be avoided.

8. With reference to main reduction gearing explain why:

(a)   nodal drives are usually associated with high pressure turbines,
(b)   pinions are generally of alloy steel and wheel rims are not,
(c)   pitting is usually found at the ends of the teeth.

9. Draw diagrammatically a set of epicyclic/helical main gearing associated with a set of cross compound turbines, labelling the principal items and showing the direction of rotation of all shafts. Describe the arrangement as drawn. Explain why some reductions are epicyclic and others are not.

10. With reference to main reduction gearing explain the reason for:

(a) tooth form,
(b) surface pitting,
(c) single plane configuration.

11. With reference to main gearing explain:

(a)   how progressive pitting may be arrested,
(b)   how vapour space corrosion is kept to a minimum and why this is necessary,
(c)   why records are kept.

12. With reference to main reduction gearing:

(a)   explain the purpose of quill shafts,
(b)   sketch and describe a quill shaft making particular reference to shaft support,
(c)   state how quill shafts are usually connected to the turbine rotors and gear pinions.

13. With reference to main reduction gearing:

(a)   explain why helical tooth formation is invariably used,
(b)   give two reasons why double helical gearing is frequently used,
(c)   state what are the advantages of single helical gearing.

14. Describe with sketches the provision for expansion of:

(a)   turbine casings,
(b)   turbine rotors,
(c)   inter-turbine eduction pipes.

15. Sketch a turbine emergency stop valve. Explain how steam is made available to the turbines when the valve has operated due to lubricating oil supply failure. State why steam supply to the turbines is required in such circumstances.

16. Draw a line diagram of a main lubricating oil system for a set of cross compound turbines, labelling all the principal components and showing the direction of flow in all lines. Give reasons why the following items are incorporated in many such systems:

(a) gravity tanks,
(b) magnetic filters.

Explain what emergency facility is inherent in 'gravity head' systems.

17. Sketch a 'gashed' turbine rotor. Show in detail how the blades are carried. Explain why most high pressure turbine rotors are 'gashed' whilst low pressure turbine rotors are often built up.

18. Draw in detail a device for protecting main turbine rotors against overspeed and excessive axial movement. Explain how it operates and protection is afforded to the engine. State how rotor overspeed and excessive axial movement can occur.

19. Draw in diagrammatic form a double casing turbine showing all connexions and directions of flow of the steam. Describe its operation. Give a good reason for this design.

20. Draw in detail a coarse tooth coupling between turbine and pinion. State what clearances are critical and how they are measured. Identify two defects of such couplings and explain how they are countered.

21. Draw in detail the devices fitted to guard against:

(a) main turbine overspeed,
(b) main lubricating oil supply failure,
(c) loss of vacuum in main condenser.

Describe how each of these devices functions.

22. Draw in detail a turbine thrust/adjusting block. Describe how it functions. Explain how rotor axial movement and location is adjusted.

23. Draw in detail a turbine diaphragm gland. Explain how it functions. Explain how and why it differs from casing glands.

24. With reference to turbine rotors explain with sketches what is meant by the terms:

(a) statically balanced,

(b) dynamically balanced.

Give two reasons why rotors are so balanced.

25. Sketch in detail an impulse turbine blade showing the configuration of the blade section and root. Similarly, draw respective diagrams for a 'reaction' turbine blade. Explain fully why they differ.

26. Explain how the tubes and tube plates of regenerative main condensers are supported. Explain how differential expansion between tubes and shell is accomodated if the tubes are rigidly held at both ends. State how condensate contamination occurs.

27. Draw in detail a feed water de-aerator, labelling the principal components and showing the direction of flow of all fluids. Explain how it works. Suggest why de-aerators supplement air ejectors in many closed feed systems.

28. Sketch and describe a main condensate extraction pump. Explain how:

(a) the pump works effectively at condenser pressure,

(b) the ingress of air through the pump is prevented.

(c) pumping continues during light load.

29. Sketch and describe a two stage air ejector. Explain how air and gases in the condenser are ejected to atmosphere. Describe the atmosphere and condenser vacuum sealing arrangements. Give a good reason why ejector efficiency 'falls off'.

N.B. It is strongly recommended that Second Class candidates attempt, in reduced detail, the First Class specimen questions.

INDEX

REED'S MARINE ENGINEERING SERIES

Vol 1 Mathematics
Vol 2 Applied Mechanics
Vol 3 Applied Heat
Vol 4 Naval Architecture
Vol 5 Ship Construction
Vol 6 Basic Electrotechnology
Vol 7 Advanced Electrotechnology
Vol 8 General Engineering Knowledge
Vol 9 Steam Engineering Knowledge
Vol 10 Instrumentation and Control Systems
Vol 11 Engineering Drawing
Vol 12 Motor Engineering Knowledge

Reed's Engineering Knowledge for Deck Officers
Reed's Maths Tables and Engineering Formulae
Reed's Marine Distance Tables
Reed's Marine Insurance
Reed's Sextant Simplified
Reed's Skipper's Handbook
Reed's Maritime Meteorology
Reed's Sea Transport – Operation and Economics

These books are obtainable from all good nautical
booksellers or direct from:

Macmillan Distribution Ltd
Brunel Road
Houndsmill
Basingstoke
RG21 6XS

Tel: 01256 302692
Fax: 01256 812558/812521

Email: mdl@macmillan.co.uk